Lecture Notes in Mathematics

Edited by A. Dold and B. Eckma...

554

Jonathan D. H. Smith

Mal'cev Varieties

Springer-Verlag
Berlin · Heidelberg · New York 1976

Author
Jonathan D. H. Smith
Department of Pure Mathematics
and Mathematical Statistics
University of Cambridge
16 Mill Lane
GB-Cambridge CB2 1SB

Library of Congress Cataloging in Publication Data

Smith, Jonathan D H 1949-
Mal'cev varieties.

(Lecture notes in mathematics ; 554)
Includes bibliographical references and indexes.
1. Algebra, Universal. 2. Abelian categories.
3. Loops (Group theory) I. Title. II. Series: Lecture notes in mathematics (Berlin) ; 554.
QA3.L28 no. 554 [QA251] 510'.8s [512] 76-50059

AMS Subject Classifications (1970): 08A05, 08A15, 08A25, 18E10, 18H99, 20N05

ISBN 3-540-07999-8 Springer-Verlag Berlin · Heidelberg · New York
ISBN 0-387-07999-8 Springer-Verlag New York · Heidelberg · Berlin

Printed in Germany
Printing and binding: Beltz Offsetdruck, Hemsbach/Bergstr.

P R E F A C E

Classes of algebras such as groups, rings, Lie algebras, and quasigroups have two kinds of property. The first kind is the particular : those properties arising from what the algebras "are", in the sense that groups "are" automorphisms of structures, rings "are" endomorphisms of abelian groups, Lie algebras "are" matrices under Lie product, and quasigroups "are" Latin squares. The second kind of property is the general, stemming from the algebraicity, the creation of new elements from old ones by operations such as multiplication and inversion. The study of the first kind of property is the task of individual theories such as ring theory or quasigroup theory. The study of the second kind of property is the task of universal algebra.

The value of universal algebra is often unclear to those working on algebras such as groups or rings which have been studied intensively for a long time. For these algebras the problems arising from the general properties have long ago been solved and taken for granted. A ring theorist knows from the beginning what a module is. Most group theorists give no second thought to the cancellation of direct products of finite groups or to the proper definition of nilpotence. But those studying less well known algebras such as quasigroups cannot afford this effete ignorance of universal algebra. If they try to beg scraps of general information from a group theorist, they will be given a horrid mixture of the particular with the general. If they then complain, they will be told that their algebras have no structure anyway. Their only recourse is to universal algebra.

However, even the group theorists and ring theorists are well advised to learn some universal algebra. Methods and results pitched at the right level of generality are forced into a lean elegance that the comparable work done under excessive assumptions may have missed. What looks messy and complicated in a particular framework may turn out to be simple and obvious in the proper general one. Furthermore, there is some beautiful and fascinating mathematics in universal algebra which collapses on narrowing down to familiar special cases.

The art of universal algebra is to pick the appropriate level of generality, which is often not the full one. The thesis of these notes is that Mal'cev varieties form an appropriate framework for many purposes. This was probably first recognised by Ore in 1936 when he showed that Mal'cev algebras have modular congruence lattices and deduced a unique factorisation theorem for direct decompositions which is false for general algebras. The crucial step came in 1954 when Mal'cev proved at a modest stroke of genius that what are now called Mal'cev varieties embrace groups, rings, modules over a ring, commutative algebras, Lie algebras, Jordan algebras, loops, quasigroups, and much more besides.

These notes set out to show some of the things that can be done with Mal'cev varieties. After an introductory chapter laying down the background and the notations the major theme of centrality is introduced in Chapter 2. Chapter 3 proves Ore's unique factorisation theorem by a Fitting's Lemma method like that used for unique factorisation theorems in group theory, and then uses centrality to improve Ore's Theorem to the standards of group theory's best corresponding result. Chapter 4 gives a cancellation theorem for direct products of finite Mal'cev algebras, and in doing so introduces the relation of central isotopy. This relation collapses to isomorphism as soon as "zeros" or "identities" appear, and so is an example of something that can only exist at sufficiently high levels of generality - at that of quasigroups rather than loops, say. Chapter 5 has another example of a similar phenomenon, which crops up in an investigation of minimal varieties and simple algebras without subalgebras. Finally, Chapter 6 deals with the cohomology of Mal'cev varieties, classifying extensions and obstructions.

There are three main areas in which the material of these notes could be developed or applied. Firstly, there is much more to be discovered about Mal'cev algebras than is presented here. Secondly, there are applications in the study of quasigroups and other scantily known algebras. Thirdly, the ideas here could be expressed in more categorical language (bearing in mind the comment in the notes to Chapter 3 on page 68).

Readers are assumed to have some acquaintance with classical universal algebra (as in [Co]) and with category theory (as in [ML]). The first chapter summarises what is assumed.

The first four chapters (except Section 3.2) formed part of a course given to graduates at Cambridge in the Lent Term 1975.

I am grateful to Darwin College, Cambridge for the Charles and Katharine Darwin Research Fellowship during the tenure of which these notes were written.

D. P. M. M. S., 16 Mill Lane, CAMBRIDGE CB2 1SB, U. K.

February 1976

Jonathan D. H. Smith.

T A B L E O F C O N T E N T S

1 Rudiments and notations

This chapter establishes the language in which these notes are written. The first section mentions the elements of classical universal algebra which will be needed. Nothing is proved there - details may be found if required in any standard text such as P. M. Cohn's [Co] . Modern universal algebra has two advantages over this classical approach : elegance, avoiding the confusion about exactly what the operations of an algebra are, and abstractness, considering algebras in any category with products, not just a category of sets. These notes do not adopt the full modern approach, partly because this is still not yet sufficiently ingrained in mathematicians to be the obvious way of conceiving things, and partly because the material here is meant to be applied to particular algebras such as quasigroups, for which purpose the classical language is more appropriate. However, it will be necessary to borrow some of the abstractness of the modern approach in talking about algebras in various categories with products. Those facts which will be used are laid down (again without proof) in the second section.

The latter half of the chapter begins the real business. The third section introduces the category of congruences on an algebra, which plays a vital role in Chapters 3 and 4. Then at last Mal'cev varieties are defined in the fourth section, and lots of familiar algebras are shown to be Mal'cev varieties.

1.1 Universal algebra.

If A and B are small sets, ${}^{B}A$ will denote the set of mappings from B to A. In particular if n is in the set $\mathbb{N}$ of natural numbers, ${}^{n}A$ will denote the direct product of n copies of A. More generally, if A is an object in a category with products, ${}^{n}A$ will denote the product of n copies of A. There are projections $\pi^{i} : {}^{n}A \to A ; (a_0,\ldots,a_{n-1}) \mapsto a_i$ and diagonal embeddings $\Delta_n : A \to {}^{n}A ; a \mapsto (a,\ldots,a)$. Δ_2 will usually just be called Δ, and $A\Delta$ will be written as $\hat{A}$, called the diagonal. Given $\theta : A \to A' ; a \mapsto a'$, write ${}^{n}\theta : {}^{n}A \to {}^{n}A' ; (a_0,\ldots,a_{n-1}) \mapsto (a'_0,\ldots,a'_{n-1})$. Given $\theta_i : A_i \to A'_i$, write $(\theta_1,\ldots,\theta_n) : A_1 \times\ldots\times A_n \to A'_1 \times\ldots\times A'_n ; (a_1,\ldots,a_n) \mapsto (a_1\theta_1,\ldots,a_n\theta_n)$.

An <u>operator domain</u> is a set Ω of <u>operations</u> together with a map $ar : \Omega \to \mathbb{N}$. If $\omega \in \Omega$, $ar(\omega)$ is called the <u>arity</u> of ω . If ω has arity $0 , 1 , 2 , 3 , \ldots , n$, it is described as <u>nullary</u>, <u>unary</u>, <u>binary</u>, <u>ternary</u>, ..., n-<u>ary</u> . An Ω-<u>algebra</u> $(A , \Omega , \rho : \Omega \to \bigcup_{n \in \mathbb{N}} {}^{({}^{n}A)}A)$ consists of a set A , the <u>underlying set</u>, an operator domain Ω , and a map $\rho : \Omega \to \bigcup_{n \in \mathbb{N}} {}^{({}^{n}A)}A$ such that $ar(\omega) = n \Rightarrow \rho(\omega) \in {}^{({}^{n}A)}A$. ρ is called the <u>action</u> of Ω on A . If there are no nullary operations, A may be empty. This possibility is usually ruled out, except at the end of Section 2.3 where it is unavoidable. Ω and ρ may be omitted from the notation for an Ω-algebra if they are clear from the context. In particular, it will be useful (except in parts of Chapter 4) to confuse an algebra with its underlying set. The image of an n-tuple $(a_1 , \ldots , a_n)$ of elements of A under $\rho(\omega)$ for n-ary ω will be written variously as

$$(a_1 , \ldots , a_n)\rho(\omega) = a_1 \ldots a_n\rho(\omega) = a_i\rho(\omega)$$
$$= (a_1 , \ldots , a_n)\omega = a_1 \ldots a_n\omega = a_i\omega .$$

Other notations may also be used, such as $a + a'$ for a binary operation $+$. Ω-algebras, with the homomorphisms between them, form a category. Isomorphism is denoted by $\cong$. The one-element algebra $\{1\}$ or 1 is a terminal object, and $\emptyset$ or 1 is initial.

Let X be a set. Then an Ω-<u>word</u> in X (or just <u>word</u>) is an element of X or, inductively, $w_1 \ldots w_n\omega$, where ω is an n-ary operation of Ω and the w_i are words. (Note that if η is a nullary operation, η is a word.) Let w be an Ω-word in X, and let $x_1, \ldots, x_k$ be all the elements of X appearing in it. Given an Ω-algebra (A,Ω,ρ), and an Ω-word w in X involving the elements $x_1, \ldots, x_k$ of X, w and ρ determine a mapping from ${}^kA$ to A sending $(a_1,\ldots,a_k)$ to the element of A obtained by replacing any x_i appearing in w by a_i and any ω appearing in w by $\rho(\omega)$. In this way the word w may be assigned the arity k and adjoined to Ω. "Operation" will variously denote either the original operations or the words regarded as operations.

Let w_1 and w_2 be operations in Ω. An Ω-algebra (A,Ω,ρ) is said to satisfy the identical relation $w_1 = w_2$ if $\rho(w_1) = \rho(w_2)$. The pair (w_1,w_2) is called an identity. If S is such a set of identities, then the class of all Ω-algebras in which all the identities S are satisfied is called a variety of Ω-algebras. Varieties form full subcategories of the category of Ω-algebras.

If (A,ρ) is an Ω-algebra, then ${}^2A$ with the pointwise actions is also an Ω-algebra $({}^2A, {}^2\rho)$. If (B,σ) is a further Ω-algebra, $A \times B$ with the action ρ on the first component and σ on the second is denoted by $(A \times B, (\rho,\sigma))$ and called the *direct product* of (A,ρ) and (B,σ) . Let C be a subset of A (written $C \subseteq A$) . If C is closed under the action ρ , (C,ρ) is said to be a *subalgebra* of (A,ρ) , written $(C,\rho) \leqslant (A,\rho)$. Proper subset and subalgebra relations will be denoted by $\subset$ and $<$ respectively. If S is a subset of A , let $< S >$ denote the least (under $\leqslant$) subalgebra of (A,ρ) containing S , the subalgebra *generated by* S . Let V be an equivalence relation on A , i.e. a reflexive, symmetric, and transitive subset of ${}^2A$. The notation $x \; V \; y$ will often be used instead of $(x,y) \in V$, as well as the pictorial notation

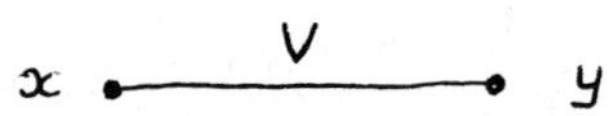

This pictorial notation is developed later with a concept of "completing parallelograms", and is a useful source of intuition. The *equivalence class* x^V is $\{ y \in A \mid x \; V \; y \}$. A^V will denote the set of equivalence classes, and $\mathrm{nat}V : A \to A^V \; ; \; x \mapsto x^V$ the *natural projection*. If V is also a subalgebra of ${}^2A$, it is called a *congruence* on A . Then (A^V,ρ) with $\rho(\omega) : (a_1^V,\ldots,a_n^V) \mapsto a_1\ldots a_n\rho(\omega)^V$ forms an Ω-algebra called the *quotient* of A by V . If $f : (A,\rho) \to (B,\sigma)$ is a homomorphism of Ω-algebras, $\ker f = \{ (x,y) \in {}^2A \mid xf = yf \}$ is a congruence on A , called the *kernel* of f , with $A \;\mathrm{nat}\ker f = Af$. Full subcategories of the category of Ω-algebras are varieties iff they are closed under the taking of direct products, subalgebras, and quotients.

The set $X\Omega$ of Ω-words in a set X forms an Ω-algebra under the action ρ given by $(w_1,\ldots,w_n)\rho(\omega) = w_1\ldots w_n\omega$. If S is the set of identities in X satisfied by a variety $\underline{\underline{T}}$ of Ω-algebras, S is a congruence on X. The quotient $(X\Omega^S,\rho)$ is called the <u>free</u> $\underline{\underline{T}}$-algebra $F_{\underline{\underline{T}}}(X)$ on the set X. If (Y,σ) is in $\underline{\underline{T}}$, any set-mapping $f : X \to Y$ has a unique extension to a $\underline{\underline{T}}$-morphism $\overline{f} : (X\Omega^S,\rho) \to (Y,\sigma)$ with the restriction $\overline{f}|_X = f$.

1.2 Algebras in categories.

Let $\underline{\underline{T}}$ be a variety of Ω-algebras satisfying the identities S, and let $\underline{\underline{C}}$ be a category with products. Then a <u>$\underline{\underline{T}}$-algebra in $\underline{\underline{C}}$</u> is an object C of $\underline{\underline{C}}$ together with a $\underline{\underline{C}}$-morphism $f_\omega : {}^{ar(\omega)}C \to C$ for each operation ω in Ω such that if (w,w') is in S, then $f_w = f_{w'}$. A <u>morphism of $\underline{\underline{T}}$-algebras in $\underline{\underline{C}}$</u> is a $\underline{\underline{C}}$-morphism $e : C \to D$ such that all the diagrams

$$\begin{array}{ccc} {}^{n}C & \xrightarrow{\;{}^{n}\theta\;} & {}^{n}D \\ {\scriptstyle f_\omega}\big\downarrow & & \big\downarrow{\scriptstyle f_\omega} \\ C & \xrightarrow{\;\theta\;} & D \end{array}$$

commute. Thus the $\mathbf{T}$-algebras in $\mathbf{C}$ form a category $\mathbf{T} \boxtimes \mathbf{C}$. For example, if $\mathbf{Set}_0$ denotes the variety of <u>pointed sets</u>, i.e. sets with one nullary operation, $\mathbf{T}_0 = \mathbf{Set}_0 \boxtimes \mathbf{T}$ is the category of <u>pointed $\mathbf{T}$-algebras</u>. A $\mathbf{T}_0$-algebra is a $\mathbf{T}$-algebra A together with a $\mathbf{T}$-morphism $\{1\} = {}^{0}A \to A$, i.e. a selected singleton subalgebra $\{0_A\} \leqslant A$. A $\mathbf{T}_0$-morphism $\theta : A \to B$ is a $\mathbf{T}$-morphism for which $0_A\theta = 0_B$. Note that if $\mathbf{Set}$ denotes the variety of (small) sets, then $\mathbf{Set} \boxtimes \mathbf{T} = \mathbf{T} \boxtimes \mathbf{Set} = \mathbf{T}$.

Let C be a $\mathbf{T}$-algebra in $\mathbf{C}$, B an object of $\mathbf{C}$, and let $\mathbf{C}(B,C)$ denote the set of $\mathbf{C}$-morphisms from B to C. Let ω be an n-ary operation of Ω, let $1 \in {}^{n}\mathbf{C}(B,C)$, and let $\varphi : {}^{n}\mathbf{C}(B,C) \to \mathbf{C}(B,{}^{n}C)$ be the natural isomorphism coming from the definition of the product ${}^{n}C$. Then the definition $\rho(\omega) : {}^{n}\mathbf{C}(B,C) \to \mathbf{C}(B,C)\ ;\ 1 \mapsto 1\varphi f_\omega$ makes $(\mathbf{C}(B,C),\Omega,\rho)$ into a $\mathbf{T}$-algebra. Conversely, if C is an object of $\mathbf{C}$ such that $(\mathbf{C}(B,C),\Omega,\rho)$ is a $\mathbf{T}$-algebra for each object B of $\mathbf{C}$, let ω be an n-ary operation. For $0 \leqslant i < n$, the projections $\pi^i : C \to C$ are in the $\mathbf{T}$-algebra $(\mathbf{C}({}^{n}C,C),\Omega,\rho)$, and then defining

$f_\omega = \pi^0 \ldots \pi^{n-1}\rho(\omega) \in \underline{\underline{C}}({}^nC,C)$ makes C into a $\underline{\underline{T}}$-algebra in $\underline{\underline{C}}$.

The category $\underline{\underline{T}}$ itself has products, and it is interesting to consider abelian groups in $\underline{\underline{T}}$. Let A be such an object, i.e. a $\underline{\underline{T}}$-algebra with $\underline{\underline{T}}$-morphisms $+ , - : {}^2A \to A$ and a singleton subalgebra $\{0\} \leqslant A$ satisfying the appropriate axioms. Let ω in Ω be an n-ary operation. Then for $x_1 , \ldots , x_n$ in A ,

$$\begin{aligned}(x_1,x_2,\ldots,x_n)\omega &= (x_1+0,0+x_2,\ldots,0+x_n)\omega \\ &= (x_1,0,\ldots,0)\omega + (0,x_2,\ldots,x_n)\omega \\ &= \ldots \\ &= (x_1,0,\ldots,0)\omega + (0,x_2,\ldots,0)\omega + (0,0,\ldots,x_n)\omega .\end{aligned}$$

Defining $t_{i\omega} : A \to A \; ; \; x \mapsto (0,\ldots,\underset{\substack{\uparrow \\ \text{i-th component}}}{x},\ldots,0)\omega$,

$(x_1,\ldots,x_n)\omega = x_1 t_{1\omega} + \ldots + x_n t_{n\omega}$. Let T be the set of all such mappings, which are known as <u>translations</u>. T has a multiplication defined by composition, and an addition defined by $x(t+t') = xt+xt'$. T becomes a ring, and A a T-module. The structure of A as an abelian group in $\underline{\underline{T}}$, in particular as a $\underline{\underline{T}}$-algebra, is specified precisely by its structure as a T-module.

Let R be a $\underline{\underline{T}}$-algebra. The category of <u>$\underline{\underline{T}}$-algebras under</u> R , $R/\underline{\underline{T}}$, has as objects all $\underline{\underline{T}}$-morphisms $R \to T$, and as morphisms all commutative diagrams

$$\begin{array}{ccc} & R & \\ \swarrow & & \searrow \\ T & \longrightarrow & T' \end{array} \quad .$$

The category of $\underline{\underline{T}}$-*algebras over* R , $\underline{\underline{T}}/R$, has as objects all $\underline{\underline{T}}$-morphisms $T \to R$, and as morphisms all commutative diagrams

$$\begin{array}{ccc} T & \xrightarrow{f} & T' \\ \searrow & & \swarrow \\ & R & \end{array} \quad .$$

Sometimes one just refers to $f : T \to T'$. Note that $\underline{\underline{T}}/\{1\}$ may be considered just as $\underline{\underline{T}}$. Algebras under R are its quotients, and algebras over R its *extensions*.

The category $\underline{\underline{T}}/R$ has products : given $f : T \to R$ and $f' : T' \to R$, form the pullback of

$$\begin{array}{ccc} & & T' \\ & & \downarrow \\ T & \longrightarrow & R \end{array}$$

in $\underline{\underline{T}}$. This may be taken to be $T \times_R T' = \{ \, (t,t') \in T \times T' \mid tf = t'f' \, \}$, with $T \times_R T' \to R \, ; \, (t,t') \mapsto tf = t'f'$. Then the universality property of pullbacks says that $T \times_R T' \to R$ is the product of $T \to R$ and $T' \to R$ in $\underline{\underline{T}}/R$.

The category of abelian groups in $\underline{\underline{T}}/R$ is called the *category of* R-*modules*. If $T \to R$ is an R-module, and $S \to R$ is an object of $\underline{\underline{T}}/R$, then as above $\underline{\underline{T}}/R(S \to R, T \to R)$ is an abelian group. In particular $\underline{\underline{T}}/R(\, 1_R : R \to R \, , \, T \to R \,)$ is called the group $\mathrm{Der}(R,T)$ of *derivations from* R *to* T . If $T \to R$ is an R-module and $S \to R$

is a $\underline{\underline{T}}$-morphism, then the pullback
$$\begin{array}{ccc} S \times_R T & \longrightarrow & T \\ \downarrow & & \downarrow \\ S & \longrightarrow & R \end{array}$$

produces an S-module $S \times_R T \to S$.

1.3 <u>The category of congruences.</u>

If U and V are equivalence relations on a set A , let $U \circ V = \{ (x,y) \in {}^2A \mid \exists\, t \in A .\ x\,U\,t\,V\,y \}$ and $U \vee V =$
$\{ (x,y) \in {}^2A \mid \exists\, n \in \mathbb{N} .\ \exists\, t_0 , t_1 ,\ldots, t_{2n} \in A .$
$$x = t_0\,U\,t_1\,V\,t_2\,U \ldots U\,t_{2n-1}\,V\,t_{2n} = y \} .$$

$U \circ V$:

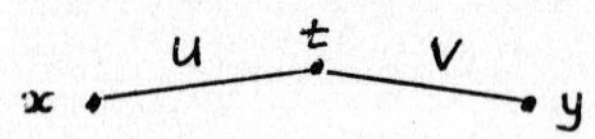

$U \vee V$:

$U \vee V$ is an equivalence relation on A called the <u>join</u> of U and V . If the equivalence relations on A are partially ordered by the inclusion relation, the join of U and V is the least upper bound of U and V . The set-theoretic intersection $U \cap V$ of U and V is also an equivalence relation on A , the greatest lower bound of U and V .

If A is an Ω-algebra and U , V are congruences on A , then $U \circ V$, $U \vee V$, and $U \wedge V$ are also subalgebras of ${}^2A$. Thus $U \vee V$ and $U \wedge V$ are congruences on A . In this way the congruences on A form a lattice. The lattice may be viewed as a category by seeing an arrow $f : U \to V$ whenever $U \leqslant V$.

For many purposes, such as the study of direct decompositions in Chapter 3, the lattice of congruences does not have enough arrows, so it is necessary to introduce some more. If U is a congruence on A , then it is a reflexive subalgebra of ${}^2A$, and so there are homomorphisms

$$A \xrightarrow{\Delta} U \xrightarrow{\pi^0} A$$

of Ω-algebras. A <u>morphism of congruences</u> is a homomorphism $\varphi : U \to V$ of Ω-algebras from the congruence U to the congruence V such that the diagram

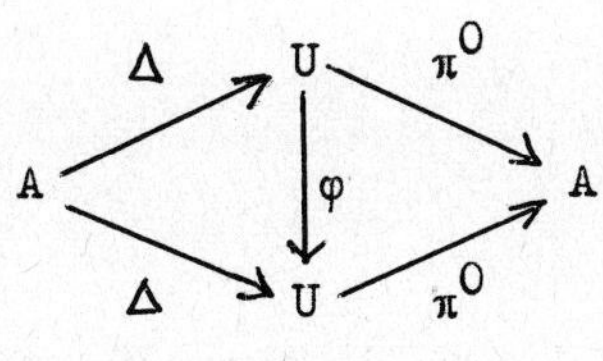

commutes. In the pictorial notation, $\varphi : U \to V \; ; \; (x,y) \mapsto (x,z)$ may be thought of as a pivoting about x :

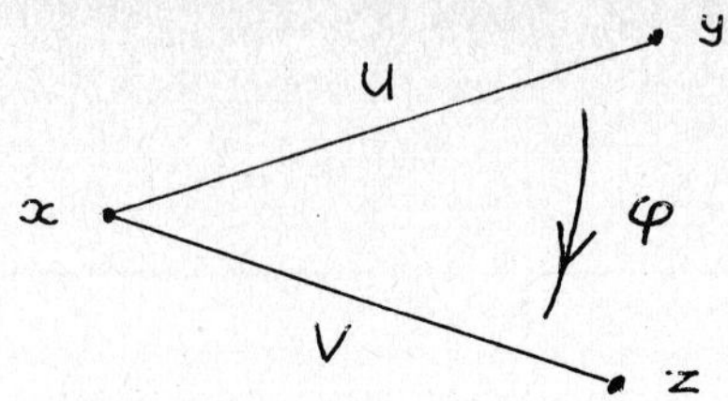

If $U \leqslant V$, then the injection of U in V is a morphism of congruences. The congruences on A with the morphisms of congruences between them form the <u>category of congruences</u> on A . It has the lattice of congruences on A as a subcategory. Note that $\hat{A}$ is a null object (i.e. both initial and terminal) in the category of congruences.

If U and V are equivalence relations on A for which $U \circ V = V \circ U$, then $U \circ V = U \vee V$:

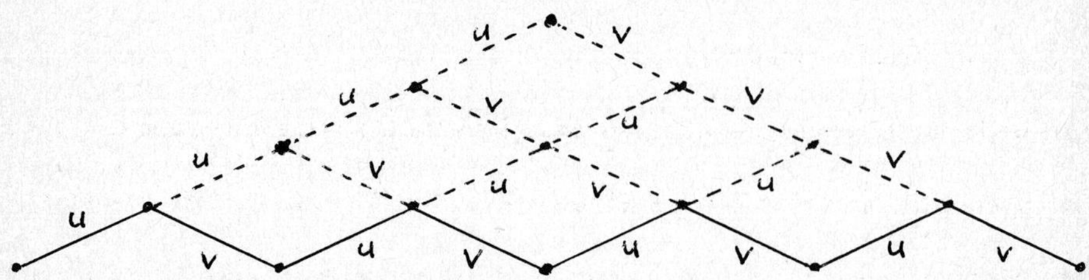

U and V are said <u>to commute</u>. If U and V are congruences on the algebra A which commute and for which $U \cap V = \hat{A}$, then the join $U \circ V = U \vee V$ of U and V is called the <u>direct product</u> $U \sqcap V$ of U and V . The reason for this nomenclature is given in Proposition 215:

if $U \sqcap V$ exists (i.e. if $U \cap V = \hat{A}$ and $U \circ V = V \circ U$), then $U \sqcap V$ is the product of U and V in the category of congruences on A .

Let W be a congruence contained in $U \sqcap V$. If $(x,y) \in W$, $(x,y) \in U \sqcap V$, so $\exists\, z \in A$. $x\, U\, z\, V\, y$. z is the unique such, since $x\, U\, z'\, V\, y \;\Rightarrow\; (z,z') \in U \cap V = \hat{A}$. This is written : $\exists_1\, z \in A$. $x\, U\, z\, V\, y$.

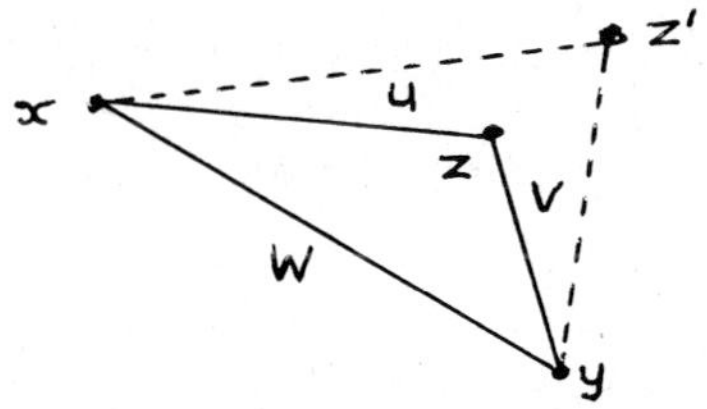

The mapping $f : W \to U\, ;\, (x,y) \mapsto (x,z)$ is a morphism of congruences, called the <u>projection of</u> W <u>onto</u> U <u>along</u> V . $\{ (x,z) \mid \exists\, (x,y) \in W .\; x\, U\, z\, V\, y \}$ is also called the <u>projection of</u> W <u>onto</u> U <u>along</u> V , and denoted by U_V^W . If $U \sqcap V = W \sqcap V$, then the projection of U onto W along V is a two-sided inverse for the projection of W onto U along V , so U and W are isomorphic in the category of congruences on A . This isomorphism is examined more closely under certain conditions in Proposition 322.

For the rest of this section, suppose that the congruences on A all commute. Then the direct product is very well behaved. For example, it is associative in the sense that if

$U \sqcap (V \sqcap W)$ exists, then $(U \sqcap V) \sqcap W$ also exists and equals $U \sqcap (V \sqcap W)$. For $U \cap (V \circ W) = \hat{A}$ and $V \cap W = \hat{A}$. Clearly $U \cap V = \hat{A}$. If $(x,y) \in W \cap (U \sqcap V)$, say $x\ U\ t\ V\ y$, then $(x,t) \in U \cap (V \circ W) = \hat{A}$, so $(x,y) \in V \cap W = \hat{A}$. For this reason

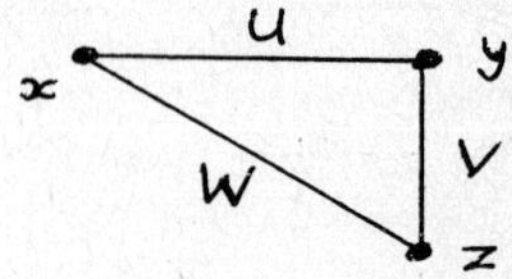

multiple direct products are written without brackets. Note that $\sqcap$ always commutes in the sense that if $U \sqcap V$ exists, then $V \sqcap U$ exists and equals $U \sqcap V$.

A congruence U is said to be (<u>directly</u>) <u>indecomposable</u> if $U > \hat{A}$, but $U = U' \sqcap U''$ implies $U' = \hat{A}$ or $U'' = \hat{A}$.

131 PROPOSITION If U is a congruence on A with the minimal condition on subcongruences, then U can be expressed as a direct product of indecomposables.

<u>Proof.</u> If U is indecomposable, there is nothing to prove. Otherwise, $U = U_1 \sqcap U_2$ with $U > U_1 > \hat{A}$. If U_1 is indecomposable, it is the first sought factor found. If not, $U_1 = U_1' \sqcap U_1''$, and $U = U_1' \sqcap U_1'' \sqcap U_2$, with $U > U_1 > U_1' > \hat{A}$. Continue thus. The descending chain $U > U_1 > U_1' > \dots$ stops, at the first indecomposable direct factor. Call it V_1. Then $U = V_1 \sqcap U_2'$, say. Now U_2' has the minimal condition on subcongruences, and so can be

expressed as $U'_2 = V_2 \sqcap U'_3$, V_2 being indecomposable. Then $U = V_1 \sqcap V_2 \sqcap U'_3$. Continue thus. The descending chain $U > U'_2 > U'_3 > \dots$ stops. Thus the whole process does, and the required expression is found.]

If the congruences on A all commute, then the lattice of congruences on A satisfies the <u>modular law</u>

$$U \geqslant V \Rightarrow U \cap (V \circ W) = V \circ (U \cap W) .$$

Note that, always, $V \vee (U \cap W) \leqslant U \cap (V \vee W)$. For the converse, let $(x,y) \in U \cap (V \circ W)$, say $x \; U \; y$, $x \; V \; t \; W \; y$:-

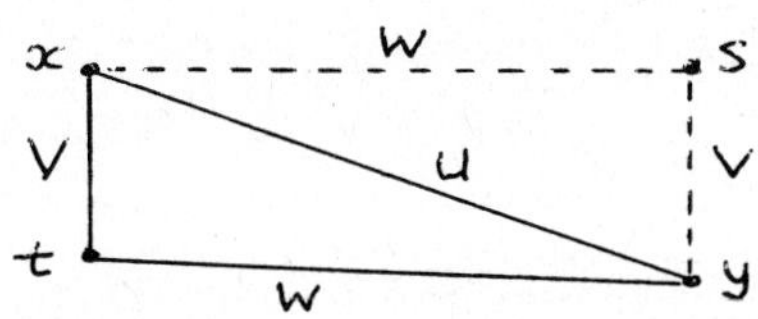

.

Then $\exists \; s \in A \, . \; x \; W \; s \; V \; y$. But $V \leqslant U$, so $s \; U \; x$. Then $y \; V \; s \; (U \cap W) \; x$, whence $(x,y) \in V \circ (U \cap W)$, as required.

In modular lattices there is a result known as <u>Ore's Theorem</u> [Bi, Ch. VII, §7] which for the lattice of congruences on the algebra A becomes :

132 THEOREM If U is a congruence on A with the maximal and minimal conditions on subcongruences, then its expression as a direct

product of indecomposables given by Proposition 131 is unique.

The usual lattice-theoretic proof of Ore's Theorem proceeds by induction and analysis of different cases. In Section 3.1 a different proof of Theorem 132 is given, analogous to that of unique factorisation theorems in group theory and lending itself better to the weakening of hypotheses that takes place in Section 3.2.

If $^2A = U \sqcap V$, then every element of A is uniquely specified by its U-class and its V-class. Thus $^2A \cong A^U \times A^V$. Conversely, if $\theta : A \to B \times C$ is an isomorphism, and $\varphi : B \times C \to B$, $\psi : B \times C \to C$ are the projections, let $U = \ker\theta\varphi$, $V = \ker\theta\psi$. Then $^2A = U \sqcap V$.

If $^2A = U \sqcap V$, V is called the (<u>direct</u>) <u>complement</u> of U , sometimes written $\overline{U}$. This notation is only used if a unique such V has been specified. Generalising the above, if $^2A = U_1 \sqcap U_2 \sqcap \cdots \sqcap U_n$, then $A \cong \mathrm{Anat}\overline{U}_1 \times \mathrm{Anat}\overline{U}_2 \times \ldots \times \mathrm{Anat}\overline{U}_n$.

An algebra B is said to be <u>indecomposable</u> if $|B|$, the number of elements in (the underlying set of) B , is greater than 1 , but $B \cong C \times D$ implies $C = \{1\}$ or $D = \{1\}$. If $^2A = U_1 \sqcap U_2 \sqcap \cdots \sqcap U_n$, then $\mathrm{Anat}\overline{U}_1$ is indecomposable (as an algebra) iff U_1 is indecomposable (as a congruence). For if $U_1 = U' \sqcap U''$, $\mathrm{Anat}\overline{U}_1 \cong \mathrm{Anat}\overline{U}' \times \mathrm{Anat}\overline{U}''$ (where for example $\overline{U}' = U'' \sqcap U_2 \sqcap \cdots \sqcap U_n$, in accordance with the convention above), and conversely if $\mathrm{Anat}\overline{U}_1 \cong B \times C$, then U_1 is the direct product of the kernels of the projections of A onto $B \times \mathrm{Anat}\overline{U}_2 \times \ldots \times \mathrm{Anat}\overline{U}_n$ and onto $C \times \mathrm{Anat}\overline{U}_2 \times \ldots \times \mathrm{Anat}\overline{U}_n$.

1.4 Mal'cev varieties.

As Section 1.3 has shown, algebras all of whose congruences commute have several desirable properties. Even more useful are varieties consisting entirely of algebras all of whose congruences commute.

141 DEFINITION Let $\underline{\underline{T}}$ be a variety of Ω-algebras such that each pair of congruences on each $\underline{\underline{T}}$-algebra A commutes. Then $\underline{\underline{T}}$ is said to be a Mal'cev variety, and $\underline{\underline{T}}$-algebras are called Mal'cev algebras.

The name comes from the Russian algebraist A. I.

Mal'cev, who proved [**Ma**] the following innocuous-looking but far-reaching theorem on which almost everything in these notes depends :

142 THEOREM Let $\underline{\underline{T}}$ be a variety of Ω-algebras satisfying the identities S . Then $\underline{\underline{T}}$ is a Mal'cev variety iff there is a ternary operation $(x,y,z)P$ in Ω for which the identical relations $(x,y,y)P = x$ and $(x,x,z)P = z$ are satisfied.

<u>Proof.</u> Firstly, suppose that there is such an operation P . Let U , V be congruences on the $\underline{\underline{T}}$-algebra A , and let $x\ U\ y\ V\ z$.

Now $x\ V\ x$ since V is reflexive,

$y\ V\ z$ is given, and

$z\ V\ z$ since V is reflexive.

Then $(x,y,z)P\ V\ (x,z,z)P = x$, since $V \leqslant {}^2A$.

Similarly $x\ U\ x$,

$x\ U\ y$, and

$z\ U\ z$.

Then $z = (x,x,z)P\ U\ (x,y,z)P$.

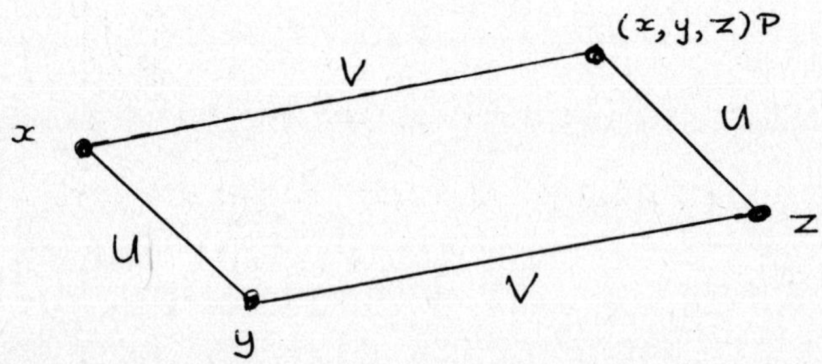

Thus $x\ U\ y\ V\ z = x\ V\ (x,y,z)P\ U\ z$, i.e. $U \circ V \subseteq V \circ U$, whence by symmetry U and V commute.

Conversely, let $\underline{\underline{T}}$ be a Mal'cev variety. Let A be the free $\underline{\underline{T}}$-algebra on the set $\{x,y,z\}$. Let U be the least congruence on A containing (x,y) , and let V be the least congruence on A containing (y,z) . Then $x\ U\ y\ V\ z$. Since U and V commute, there is an element $(x,y,z)P$ of A such that $x\ V\ (x,y,z)P\ U\ z$. Then $x^V = (x^V,y^V,y^V)P$ and $z^U = (x^U,x^U,z^U)P$. But A^U and A^V are free $\underline{\underline{T}}$-algebras on the sets $\{x^U,z^U\}$ and $\{x^V,y^V\}$ respectively. Thus $x = (x,y,y)P$ and $z = (x,x,z)P$ are identities.]

P is known as the <u>Mal'cev operation</u>.

The first consequence of Theorem 142 is the following property of Mal'cev varieties, which is so useful that it is rarely mentioned when it is applied :

143 PROPOSITION If $\underline{\underline{T}}$ is a Mal'cev variety, then for each $\underline{\underline{T}}$-algebra A , each subalgebra U of ${}^2A$ containing $\hat{A}$ is a congruence on A .

<u>Proof.</u> Let $\hat{A} \leqslant U \leqslant {}^2A$. Symmetry and transitivity of U must be proved. Let (x,y) , $(y,z) \in U$.

Now $\quad x \, U \, x$ since $\hat{A} \leqslant U$,

$\quad x \, U \, y$ is given, and

$\quad y \, U \, y$ since $\hat{A} \leqslant U$.

Then $y = (x,x,y)P \; U \; (x,y,y)P = x$, so that U is symmetric.

Also $\quad x \, U \, y$ is given,

$\quad y \, U \, y$ since $\hat{A} \leqslant U$, and

$\quad y \, U \, z$ is given.

Then $x = (x,y,y)P \; U \; (y,y,z)P = z$, so that U is transitive.]

Thus the congruences on a Mal'cev algebra A are precisely the interalgebras $\hat{A} \leqslant U \leqslant {}^2A$. As an example of the use of this, let $\varphi : U \to V$ be a morphism in the category of congruences on A . Let $\mathrm{Im}\varphi = \{ (x,z) \in V \mid \exists\, (x,y) \in U \,.\, (x,y)\varphi = (x,z) \}$ and $\mathrm{Ker}\varphi = \{ (x,y) \in U \mid (x,y)\varphi = (x,x) \}$. Then $\hat{A} \leqslant \mathrm{Im}\varphi \leqslant {}^2A$ and $\hat{A} \leqslant \mathrm{Ker}\varphi \leqslant {}^2A$, so that $\mathrm{Im}\varphi$ and $\mathrm{Ker}\varphi$ are congruences on A . (Note the use of capital letters in this notation, particularly in $\mathrm{Ker}\varphi$ to distinguish it from $\ker\varphi = \{ ((x,y),(x',y')) \in {}^2U \mid (x,y)\varphi = (x',y')\varphi \}$.) Note that φ is an epimorphism iff $\mathrm{Im}\varphi = V$, and a monomorphism iff $\mathrm{Ker}\varphi = \hat{A}$. The embedding $\mathrm{Ker}\varphi \to U$ has the usual universality property for kernels : it is the equaliser of $\varphi : U \to V$ and $U \to \hat{A} \to V$. $\mathrm{Im}\varphi$ and $\mathrm{Ker}\varphi$ are used in Section 3.1.

Mal'cev's Theorem 142 enables one to see readily that many familiar varieties are Mal'cev varieties. Taking $(x,y,z)P = xy^{-1}z$ shows that groups form a Mal'cev variety, whence abelian groups, rings, modules over a ring, commutative algebras, Lie algebras, Jordan algebras, and so on do.

More generally, quasigroups form a Mal'cev variety. A <u>quasigroup</u> Q has operations called <u>multiplication</u> $.$, <u>right division</u> $/$, and <u>left division</u> $\backslash$, satisfying the identities $(x.y)/y = x$, $(x/y).y = x$, $y\backslash(y.x) = x$, and $y.(y\backslash x) = x$. In other words, in the equation $x.y = z$, knowledge of any two of x , y , z specifies the third uniquely. Then $(x,y,z)P = (x/(y\backslash y)).(y\backslash z)$ is a Mal'cev operation for quasigroups. Much of the work of these notes was done with the application to quasigroups in mind.

Note that a finite quasigroup can be described completely by its multiplication table, which is a latin square, and conversely latin squares are the multiplication tables of quasigroups. Non-empty quasigroups with associative multiplication are groups.

NOTES

Section 1.2 is based on P. Freyd's readable paper [Fr]. Full accounts of the modern approach to universal algebra are given by F. E. J. Linton [Ec, pp. 7 - 74] and (more legibly) by G. C. Wraith [Wr].

Proposition 131 is taken from [AC, (VI,4,6), Théorème 2].

The terminology of Section 1.4, and indeed of the title of these notes, is not yet standard. The words "variety of algebras such that each pair of congruences on each algebra commutes" or even "algebras with commuting congruences" are such mouthfuls for such a useful concept that something shorter is required. Besides, these terms give a misleading emphasis to what is but one aspect. More positively, Mal'cev's contribution to the subject, Theorem 142, is so pervasive despite its modest appearance that it deserves greater recognition.

2 Centrality

Acabo siempre aludiendo al centro sin
la menor garantía de saber lo que digo,
cedo a la trampa fácil de la geometría
con que pretende ordenarse nuestra vida
de occidentales : Eje, centro, razón
de ser, Omphalos, nombres de la
nostalgia indoeuropeo.

JULIO CORTAZAR, Rayuela.

This chapter introduces the theme of centrality which runs right through these notes. In group theory centrality plays a fundamental role : in defining abelian and nilpotent groups, in the theory of direct decompositions, in setting up modules, and in cohomology. It is easily expressed there as the multiplication of certain factors being commutative. Similarly in Lie algebras and ring theory it appears as the (Lie- in the case of Lie algebras) multiplication of certain factors being zero. The two ideas which have to be combined are the independence of direct factors and the concept of (abelian) group in appropriate categories. For general algebras the basic definition is given right at the start of the first section, and the rest of that section justifies the various conditions of the definition by showing how they fit together to give the desired properties. The crucial second section deals with centrality in Mal'cev varieties, and as it does so it becomes apparent that the notion of Mal'cev variety is a good one. The last section defines nilpotence in Mal'cev varieties, and gives one brief application to Frattini subalgebras.

2.1 Centrality in general.

For this section, let $\underline{\underline{T}}$ be any variety, not necessarily Mal'cev.

211 DEFINITIONS Let A be a $\underline{\underline{T}}$-algebra, let β, γ be congruences on A, and let $(\gamma|\beta)$ be a congruence on β. Then γ is said to centralise β by means of the centreing congruence $(\gamma|\beta)$ iff the following conditions are satisfied :

(C0): $(x,y)\ (\gamma|\beta)\ (x',y') \Rightarrow x\ \gamma\ x'$.

(C1): $\forall\ (x,y) \in \beta\ ,\ \pi^0 : (x,y)^{(\gamma|\beta)} \to x^\gamma\ ;\ (x',y') \mapsto x'$ bijects.

(C2): The following three conditions are satisfied :

(RR): $\forall\ (x,y) \in \gamma\ ,\ (x,x)\ (\gamma|\beta)\ (y,y)$.

(RS): $(x,y)\ (\gamma|\beta)\ (x',y') \Rightarrow (y,x)\ (\gamma|\beta)\ (y',x')$.

(RT): $(x,y)\ (\gamma|\beta)\ (x',y')$ and $(y,z)\ (\gamma|\beta)\ (y',z')$

$$\Rightarrow (x,z)\ (\gamma|\beta)\ (x',z')\ .$$

Conditions (RR), (RS), and (RT) respectively are known as <u>respect for the reflexivity</u>, <u>symmetry</u>, and <u>transitivity</u> of β . (C2) is called <u>respect for equivalence</u>.

Intuitively, one thinks of the relation $(x,y)\ (\gamma|\beta)\ (x',y')$ as a parallelogram

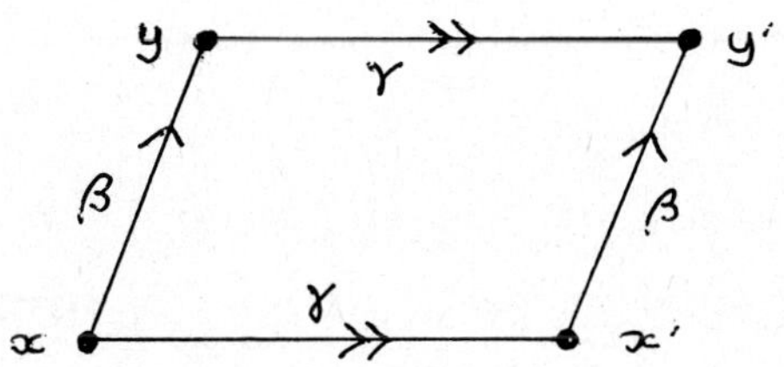

Two examples, those of groups and of (Lie) rings, should help to show how the conditions of Definition 211 work. Firstly, let A be a group, and let B , C be normal subgroups. B and C determine the congruences $\beta = \{\ (x,y) \in {}^2A \mid yx^{-1} \in B\ \}$

and $\gamma = \{ (x,y) \in {}^2A \mid yx^{-1} \in C \}$, under which they are the respective equivalence classes of the identity element. Suppose C centralises B , i.e. for all c in C and b in B , $bc = cb$. Define a relation $(\gamma|\beta)$ on β by

$$(x,y)\ (\gamma|\beta)\ (x',y') \iff \exists\, b \in B\ .\ yx^{-1} = y'x'^{-1} = b$$
$$\text{and } \exists\, c \in C\ .\ x^{-1}x' = c\ .$$

Then $(\gamma|\beta)$ is a congruence on β by means of which γ centralises β in the sense of Definition 211. For example, to check that $(\gamma|\beta)$ is a congruence on β , suppose $(x_i,y_i)\ (\gamma|\beta)\ (x_i',y_i')$ for $i = 1,2$, say $y_i x_i^{-1} = y_i' x_i'^{-1} = b_i \in B$ and $x_i^{-1}x_i' = c_i \in C$.

Then $y_1y_2(x_1x_2)^{-1} = y_1y_2x_2^{-1}x_1^{-1} = y_1b_2x_1^{-1} = b_1x_1b_2x_1^{-1}$,

and similarly $y_1'y_2'(x_1'x_2')^{-1} = b_1x_1'b_2x_1'^{-1}$.

Now $x_1'^{-1}x_1b_2x_1^{-1}x_1' = c_1^{-1}b_2c_1 = b_2$ since C centralises B .

Thus $x_1b_2x_1^{-1} = x_1'b_2x_1'^{-1}$, whence $y_1y_2(x_1x_2)^{-1} = y_1'y_2'(x_1'x_2')^{-1}$.

Also $(x_1x_2)^{-1}x_1'x_2' = x_2^{-1}x_1^{-1}x_1'x_2' = x_2^{-1}c_1x_2c_2 = c_1'c_2$

for some $c_1' = x_2^{-1}c_1x_2$ in C .

Thus $(x_1x_2,y_1y_2)\ (\gamma|\beta)\ (x_1'x_2',y_1'y_2')$.

Conversely, suppose β and γ are congruences on A , with γ centralising β by means of a centreing congruence $(\gamma|\beta)$.

Let $B = 1^\beta$, $C = 1^\gamma$. Let $b \in B$, $c \in C$.

Then $(1,b)\ (\gamma|\beta)\ (1,b)$ by reflexivity of $(\gamma|\beta)$,

$(1,1)\ (\gamma|\beta)\ (c,c)$ by (RR) for $(\gamma|\beta)$, and

$(1,b^{-1})(\gamma|\beta)(1,b^{-1})$ by reflexivity of $(\gamma|\beta)$.

Thus $(1,1)\ (\gamma|\beta)\ (c,bcb^{-1})$, since $(\gamma|\beta)$ is a congruence.

By the transitivity of $(\gamma|\beta)$, $(c,c)\ (\gamma|\beta)\ (c,bcb^{-1})$.

The condition (C1) then gives that $c = bcb^{-1}$.

Thus C centralises B in the usual group-theoretic sense.

Secondly, let A be a ring, with multiplication denoted by juxtaposition. (Alternatively, let A be a Lie ring or algebra, with Lie multiplication denoted by juxtaposition.) Let B , C be ideals of A . B and C determine the congruences $\beta = \{\ (x,y) \in {}^2A \mid y - x \in B\ \}$ and $\gamma = \{\ (x,y) \in {}^2A \mid y - x \in C\ \}$, under which they are the respective equivalence classes of the zero element. Suppose $BC + CB = 0$. Define a relation $(\gamma|\beta)$ on β by $(x,y)\ (\gamma|\beta)\ (x',y') \Leftrightarrow \exists\, b \in B\ .\ y - x = y' - x' = b$

and $\exists\, c \in C\ .\ x' - x = c$.

Then $(\gamma|\beta)$ is a congruence on β by means of which γ centralises β in the sense of Definition 211. For example, to check that $(\gamma|\beta)$ is a congruence on β , suppose $(x_i,y_i)\ (\gamma|\beta)\ (x_i',y_i')$ for $i = 1,2,$ say $y_i - x_i = y_i' - x_i' = b_i \in B$ and $x_i' - x_i = c_i \in C$.

Then $y_1'y_2' - x_1'x_2' = (x_1' + b_1)(x_2' + b_2) - x_1'x_2' = b_1b_2 + b_1x_2' + x_1'b_2 = b_1b_2 + b_1(x_2 + c_2) + (x_1 + c_1)b_2 = b_1b_2 + b_1x_2 + x_1b_2 = y_1y_2 - x_1x_2$, and $x_1'x_2' - x_1x_2 = (x_1 + c_1)(x_2 + c_2) - x_1x_2 = c_1c_2 + c_1x_2 + x_1c_2 \in C$. Thus $(x_1x_2, y_1y_2)\ (\gamma|\beta)\ (x_1'x_2', y_1'y_2')$. Conversely, suppose β and γ are congruences on A, with γ centralising β by means of a centreing congruence $(\gamma|\beta)$. Let $B = 0^\beta$, $C = 0^\gamma$. Let $b \in B$, $c \in C$. Then $(0,b)\ (\gamma|\beta)\ (0,b)$ by reflexivity of $(\gamma|\beta)$, and

$(0,0)\ (\gamma|\beta)\ (c,c)$ by (RR) for $(\gamma|\beta)$.

Thus $(0,0)\ (\gamma|\beta)\ (0,bc)$, since $(\gamma|\beta)$ is a congruence.

Condition (C1) for $(\gamma|\beta)$ then gives that $bc = 0$. Similarly, $cb = 0$, so $BC + CB = 0$.

In the two examples, then, γ centralises β if and only if C centralises B, or $BC + CB = 0$, respectively. Note that these latter conditions are symmetrical in B and C.

In fact, the symmetry noted in the examples holds in general. Suppose that γ centralises β by means of $(\gamma|\beta)$. Define a relation $(\beta|\gamma)$ on γ by

$$(x,x')\ (\beta|\gamma)\ (y,y') \Leftrightarrow (x,y)\ (\gamma|\beta)\ (x',y').$$

Let ω be an n-ary operation on A, and suppose that for $1 \leq i \leq n$, $(x_i, x_i')\ (\beta|\gamma)\ (y_i, y_i')$. Then since $(\gamma|\beta)$ is a subalgebra of $^2\beta$, $(x_i,y_i)\ (\gamma|\beta)\ (x_i',y_i')$ for $1 \leqslant i \leqslant n$ implies $(x_1..x_n\omega,y_1..y_n\omega)\ (\gamma|\beta)\ (x_1'..x_n'\omega,y_1'..y_n'\omega)$. Thus $(x_1..x_n\omega,x_1'..x_n'\omega)\ (\beta|\gamma)\ (y_1..y_n\omega,y_1'..y_n'\omega)$, i.e. $(\beta|\gamma) \leqslant {}^2\gamma$. Since $(\gamma|\beta)$ respects equivalence of β , $(\beta|\gamma)$ is an equivalence relation. Thus it is a congruence on γ . Since $(\gamma|\beta)$ is an equivalence relation, $(\beta|\gamma)$ respects equivalence of γ . Thus $(\beta|\gamma)$ satisfies condition (C2) of Definition 211. It clearly satisfies (C0) , and it satisfies (C1) since $(\gamma|\beta)$ does. Thus Definition 211 is <u>symmetrical</u> in the sense that if γ centralises β by means of $(\gamma|\beta)$, there is a naturally defined congruence $(\beta|\gamma)$ on γ by means of which β centralises γ .

Note that if γ centralises β by means of $(\gamma|\beta)$, and β' is a congruence contained in β , then γ centralises β' by means of $(\gamma|\beta') = (\gamma|\beta) \cap {}^2\beta'$. This is known as <u>centralising by restriction</u>. Note also that every congruence α on A centralises $\hat{A}$ by means of a unique congruence $(\alpha|\hat{A})$:

$$(x,x)\ (\alpha|\hat{A})\ (y,y) \Leftrightarrow x\ \alpha\ y\ .$$

If γ centralises β by means of $(\gamma|\beta)$, then by (C0), $e^i : \beta^{(\gamma|\beta)} \to A^\gamma\ ;\ (x_0,x_1)^{(\gamma|\beta)} \mapsto x_i^\gamma$ for $i = 0$, 1 is well-defined. Then (C1) implies that

$$\begin{array}{ccc} \beta & \xrightarrow{\ \pi^0\ } & A \\ {\scriptstyle \mathrm{nat}(\gamma|\beta)}\big\downarrow & & \big\downarrow{\scriptstyle \mathrm{nat}\gamma} \\ \beta^{(\gamma|\beta)} & \xrightarrow{\ e^0\ } & A^{\gamma} \end{array}$$

is a pullback (in sets, and so in $\underline{\underline{T}}$) .

If γ centralises β by means of $(\gamma|\beta)$, and $\beta \leqslant \gamma$, then for each $r \in A^{\gamma}$ there is an operation $+$ on $(e^0)^{-1}(r) = \{ (x,y)^{(\gamma|\beta)} \mid x^{\gamma} = r \}$ as follows :

213 If $y \ \gamma \ u$, define $(x,y)^{(\gamma|\beta)} + (u,v)^{(\gamma|\beta)} = (x,z)^{(\gamma|\beta)}$,

where $(y,z) \ (\gamma|\beta) \ (u,v)$.

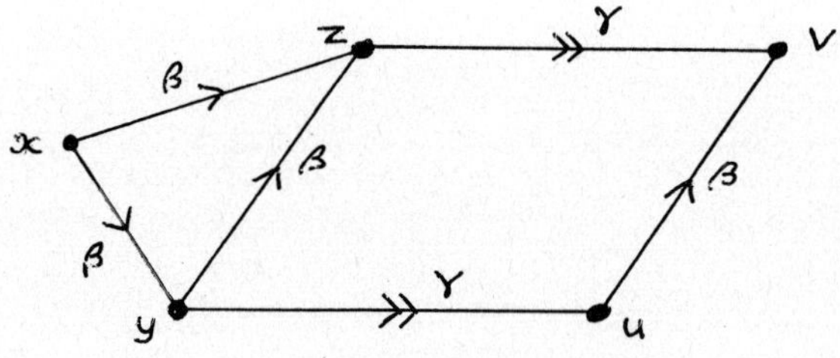

By (C1), a unique such z exists. Then by (RT) , $+$ is well-defined. Properties (C1) and (C2) imply that $((e^0)^{-1}(r),+)$ is a quasigroup. But $+$ is clearly associative, and so for each r in A^{γ} $((e^0)^{-1}(r),+)$ is a group. Summarising :

214 PROPOSITION Let A be a $\underline{\underline{T}}$-algebra, let β , γ be congruences on A with $\beta \leqslant \gamma$ and γ centralising β by means of $(\gamma|\beta)$. Then

$$\begin{array}{ccc}
\beta & \xrightarrow{\pi^{0}} & A \\
{\scriptstyle \mathrm{nat}(\gamma|\beta)}\downarrow & & \downarrow{\scriptstyle \mathrm{nat}\gamma} \\
\beta^{(\gamma|\beta)} & \xrightarrow{e^{0}} & A^{\gamma} \\
(x,y)^{(\gamma|\beta)} & \longmapsto & x^{\gamma}
\end{array}$$

is a pullback in $\underline{\underline{T}}$, and the bottom row is a group in $\underline{\underline{T}}/A^{\gamma}$.]

On a different tack, suppose that $\beta \cap \gamma = \hat{A}$ and $\beta \circ \gamma = \gamma \circ \beta$, so that the direct product $\beta \sqcap \gamma$ exists. Define $(\gamma|\beta)$ on β by $(x,y)\ (\gamma|\beta)\ (x',y') \Leftrightarrow x\ \gamma\ x'$ and $y\ \gamma\ y'$.

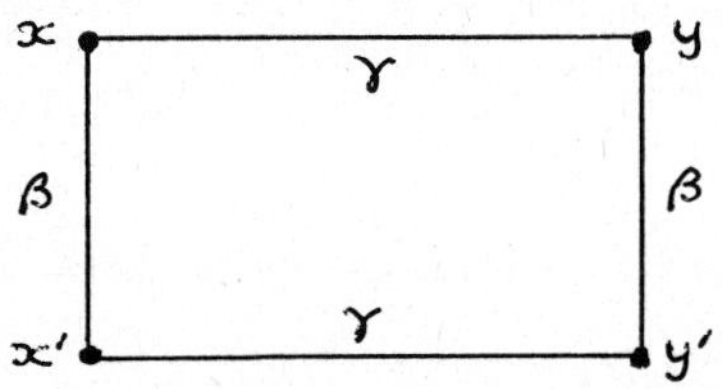

$(\gamma|\beta)$ is clearly a congruence on β , satisfying (C0) and (C2). Suppose $(x,y) \in \beta$ and $x\ \gamma\ x'$. Then $(y,x') \in \beta \circ \gamma = \gamma \circ \beta$, so $\exists\ y' \in A .\ y\ \gamma\ y'\ \beta\ x'$. Then $(x,y)\ (\gamma|\beta)\ (x',y')$. If $(x,y)\ (\gamma|\beta)\ (x,y')$, then $(y,y') \in \beta \cap \gamma = \hat{A}$. Thus $(\gamma|\beta)$ also satisfies (C1), and so is a congruence by which γ centralises β . Note that by (C0) and (RS), this $(\gamma|\beta)$ defined here is the unique centreing congruence by which γ can centralise β .

215 PROPOSITION If $\beta \sqcap \gamma$ exists, it is the product of β and γ in the category of congruences on A .

<u>Proof.</u> Define $p_\beta : \beta \sqcap \gamma \to \beta \; ; \; (a,a') \mapsto (a,b)$ and $p_\gamma : \beta \sqcap \gamma \to \gamma \; ; \; (a,a') \mapsto (a,c)$, where $a \; \beta \; b \; \gamma \; a' \; \beta \; c \; \gamma \; a$:

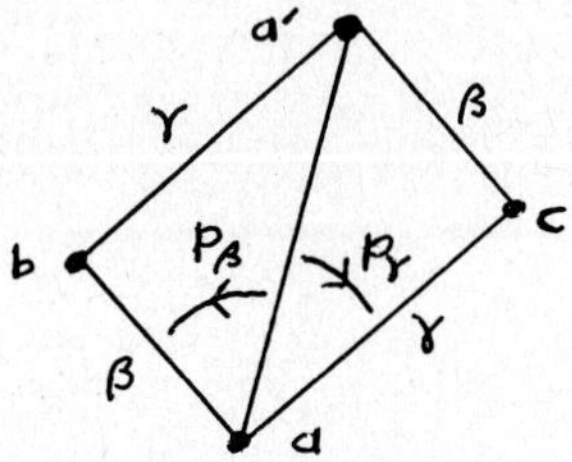

These are morphisms of congruences.

Given morphisms of congruences $f : \delta \to \beta$, $g : \delta \to \gamma$, define $f \sqcap g : \delta \to \beta \sqcap \gamma \; ; \; (a,d) \mapsto (a,a')$, where $(a,d)f = (a,b)$, $(a,d)g = (a,c)$, and $(a,b) \; (\gamma|\beta) \; (c,a')$. Then $f \sqcap g$ is a morphism of congruences, and $(f \sqcap g)p_\beta = f$, $(f \sqcap g)p_\gamma = g$. As $f \sqcap g$ is the unique morphism of congruences satisfying these latter two conditions, the result follows.]

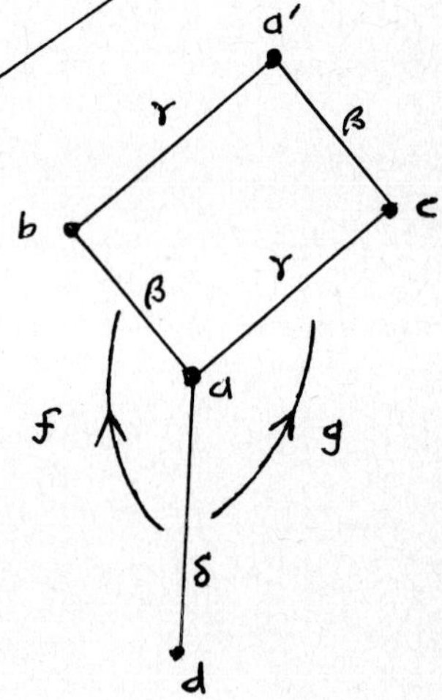

The power of the concept of centrality stems from its ability to give a unified treatment of two situations as diverse as those of Propositions 214 and 215.

If $\underline{\underline{T}}$ is not Mal'cev, γ may centralise β by means of various centreing congruences $(\gamma|\beta)$. For example, take $\underline{\underline{T}}$ to be small sets, so that congruences are just equivalence relations. Let $(G,.)$ be a not-abelian group. Let $({}^2G|{}^2G)_r$ on ${}^2G$ be the equivalence relation whose classes are just the right cosets of the subgroup $\hat{G}$, and let $({}^2G|{}^2G)_l$ on ${}^2G$ be defined similarly by the left cosets of $\hat{G}$. Then ${}^2G$ centralises itself by both of $({}^2G|{}^2G)_r$ and $({}^2G|{}^2G)_l$, but $({}^2G|{}^2G)_r$ is not equal to $({}^2G|{}^2G)_l$.

2.2 Centrality in Mal'cev varieties.

From now on, let $\underline{\underline{T}}$ be a Mal'cev variety. In Mal'cev varieties, centrality is much better behaved, intuitively because the Mal'cev operation P completes the parallelogram mentioned after Definition 211. This section begins by consecutively demonstrating three advantages :

(i) Centreing congruences are unique.

(ii) The groups of Proposition 214 are abelian.

(iii) There are necessary and sufficient conditions for centralising much briefer than those of Definition 211.

221 PROPOSITION Let A be a $\underline{\underline{T}}$-algebra, let β and γ be congruences on A, and let γ centralise β by means of $(\gamma|\beta)_1$ and $(\gamma|\beta)_2$. Then $(\gamma|\beta)_1 = (\gamma|\beta)_2$.

Proof. Let $x\ \beta\ y$ and $x\ \gamma\ x'$. Now for $i = 1, 2$,

$$(x,y)\ (\gamma|\beta)_i\ (x,y) \text{ and}$$

$$(x,x)\ (\gamma|\beta)_i\ (x,x) \text{ since } (\gamma|\beta)_i \text{ is reflexive,}$$

while $(x,x)\ (\gamma|\beta)_i\ (x',x')$ by (RR) for $(\gamma|\beta)_i$.

Hence $(x,y)\ (\gamma|\beta)_i\ (x',(y,x,x')P)$.

Thus if $(x,y)\ (\gamma|\beta)_1\ (x',y')$, then by (C1) for $(\gamma|\beta)_1$, $y' = (y,x,x')P$, and by (C1) for $(\gamma|\beta)_2$, $(x,y)\ (\gamma|\beta)_2\ (x',y')$.]

For this reason, the unique centreing congruence $(\gamma|\beta)$ is often not mentioned explicitly.

222 PROPOSITION Let A be a $\underline{\underline{T}}$-algebra, let β, γ be congruences on A with $\beta \leqslant \gamma$ and γ centralising β by means of $(\gamma|\beta)$. Then

$$\begin{array}{ccc} \beta & \xrightarrow{\pi^0} & A \\ {\scriptstyle \mathrm{nat}(\gamma|\beta)}\downarrow & & \downarrow{\scriptstyle \mathrm{nat}\gamma} \\ \beta^{(\gamma|\beta)} & \xrightarrow{e^0} & A^{\gamma} \\ (x,y)^{(\gamma|\beta)} & \longmapsto & x^{\gamma} \end{array}$$

is a pullback in $\underline{\underline{T}}$, and the bottom row is an abelian group in $\underline{\underline{T}}/A^{\gamma}$.

<u>Proof.</u> For $x \beta y$, $y \beta z$, $(y,z) (\gamma|\beta) (x,(x,y,z)P)$ and $(x,y) (\gamma|\beta) ((x,y,z)P,z)$. Then $(y,z)^{(\gamma|\beta)} + (x,y)^{(\gamma|\beta)} =$

$$(x,(x,y,z)P)^{(\gamma|\beta)} + ((x,y,z)P,z)^{(\gamma|\beta)} = (x,z)^{(\gamma|\beta)}$$

$$= (x,y)^{(\gamma|\beta)} + (y,z)^{(\gamma|\beta)} ,$$

so the operation defined by 213 is commutative.]

As a sort of converse to this proposition, if $\pi : Z \to R$ is an R-module, i.e. abelian group in $\underline{\underline{T}}/R$, define $\pi^i : \ker\pi \to Z$; $(z_0,z_1) \mapsto z_i$. Then $\ker\pi$ centralises itself by means of $\ker(\pi^0-\pi^1)$ (this is best checked with the aid of Corollary 224 below) and there is an isomorphism of $\underline{\underline{T}}$-algebras over R

$$(\ker\pi)^{\ker(\pi^0-\pi^1)} \to Z \ ; \ (z_0,z_1)^{\ker(\pi^0-\pi^1)} \mapsto z_1-z_0 \ .$$

Thus "modules are the same as self-centralising congruences".

223 PROPOSITION Let A be a $\underline{\underline{T}}$-algebra, let β , γ be congruences on A , and let $(\gamma|\beta)$ be a congruence on β . Then γ centralises β by means of $(\gamma|\beta)$ iff the following two conditions are satisfied :

(C0): $(x,y) (\gamma|\beta) (x',y') \Rightarrow x \gamma x'$.

(C3): $\forall x \in A , \ (x,x)^{(\gamma|\beta)} = (x^\gamma)\Delta$.

<u>Proof.</u> If γ centralises β by $(\gamma|\beta)$, then respect for reflexivity implies $(x^\gamma)\Delta \subseteq (x,x)^{(\gamma|\beta)}$. The opposite inclusion follows from (C1).

Conversely, suppose (C0) and (C3) are satisfied.

Condition (RR) is immediate.

Condition (C1):

Surjectivity of $\pi^0 : (x,y)^{(\gamma|\beta)} \to x^\gamma$:

Suppose $(x,y) \in \beta$ and $(x,z) \in \gamma$.

Then $(x,y)\ (\gamma|\beta)\ (x,y)$ since $(\gamma|\beta)$ is reflexive,

$(x,x)\ (\gamma|\beta)\ (x,x)$ since $(\gamma|\beta)$ is reflexive,

and $(x,x)\ (\gamma|\beta)\ (z,z)$ by (RR) .

Thus $\underline{(x,y)\ (\gamma|\beta)\ (z,(y,x,z)P)}$.

Injectivity of $\pi^0 : (x,y)^{(\gamma|\beta)} \to x^\gamma$:

Suppose $(z,z')\ (\gamma|\beta)\ (z,z'')$.

Then $(z,z)\ (\gamma|\beta)\ (z,z)$ since $(\gamma|\beta)$ is reflexive

and $(z',z)\ (\gamma|\beta)\ (z',z)$ since $(\gamma|\beta)$ is reflexive.

Hence $\underline{(z',z')\ (\gamma|\beta)\ (z',z'')}$, and so $z' = z''$ by (C3) .

Condition (RS):

Suppose $(x,y)\ (\gamma|\beta)\ (x',y')$. Then $x\ \gamma\ x'$ by (C0). Putting $z = x'$ in the calculation above verifying surjectivity of π^0 , $(x',(y,x,x')P)\ (\gamma|\beta)\ (x,y)\ (\gamma|\beta)\ (x',y')$. Thus by injectivity of π^0, $y' = (y,x,x')P$. But $(y,x,x')P\ \gamma\ y$, so $y\ \gamma\ y'$.

Now $(x,x)\ (\gamma|\beta)\ (x',x')$ by (RR),

$(x,y)\ (\gamma|\beta)\ (x',y')$ is given,

and $(y,y)\ (\gamma|\beta)\ (y',y')$ by (RR).

Hence $\underline{(y,x)\ (\gamma|\beta)\ (y',x')}$.

Condition (RT):

Suppose $(x,y)\ (\gamma|\beta)\ (x',y')$ and $(y,z)\ (\gamma|\beta)\ (y',z')$.

Then $(x,y)\ (\gamma|\beta)\ (x',y')$,

$(y,y)\ (\gamma|\beta)\ (y',y')$,

and $(y,z)\ (\gamma|\beta)\ (y',z')$.

Hence $(x,z)\ (\gamma|\beta)\ (x',z')$.]

224 COROLLARY γ centralises β by means of $(\gamma|\beta)$ iff the following two conditions are satisfied :

(C0): $(x,y)\ (\gamma|\beta)\ (x',y') \Rightarrow x\ \gamma\ x'$.

(C4): $\forall\, x \in A\ ,\ (x,x)\ (\gamma|\beta)\ (x,y) \Rightarrow x = y$.

<u>Proof.</u> Suppose conditions (C0) and (C4) are satisfied. If $x\ \gamma\ y$, then $(x,x)\ (\gamma|\beta)\ (y,(x,x,y)P) = (y,y)$, so that $(x,x)^{(\gamma|\beta)} \supseteq (x^{\gamma})\Delta$. Conversely, if $(x,x)\ (\gamma|\beta)\ (y,z)$, then by (C0) $x\ \gamma\ y$, so that $(x,x)\ (\gamma|\beta)\ (y,y)$. Transitivity of $(\gamma|\beta)$ and (C4) then imply that $y = z$, so $(x,x)^{(\gamma|\beta)} \subseteq (x^{\gamma})\Delta$. Thus (C3) is satisfied. The other implication is trivial.]

225 COROLLARY γ centralises β by means of $(\gamma|\beta)$ iff the diagram of Proposition 222 exists and is a pullback.

<u>Proof.</u> If γ does centralise β , the diagram has already been seen in Proposition 222 to exist and be a pullback. Conversely, suppose the diagram is given (hence (C0)) and is a pullback. One can

also form the pullback $B = \beta^{(\gamma|\beta)} \times_{A^\gamma} A = \{ ((x,y)^{(\gamma|\beta)},z) \mid x \gamma z \}$. Then by the uniqueness of pullbacks there is an isomorphism $\beta \to B$; $(x,y) \mapsto ((x,y)^{(\gamma|\beta)},x)$. Condition (C4) follows.]

The next result has many consequences.

226 PROPOSITION If γ centralises β_1 and β_2 , then it centralises $\beta_1 \circ \beta_2$.

Proof. Define $(\gamma|\beta_1 \circ \beta_2)$ on $\beta_1 \circ \beta_2$ by $(x,y)\ (\gamma|\beta_1 \circ \beta_2)\ (x',y')$ $\Leftrightarrow\ \exists\, t\ ,\ t'\ .\ (x,t)\ (\gamma|\beta_1)\ (x',t')$ and $(t,y)\ (\gamma|\beta_2)\ (t',y')$. Let ω be an n-ary operation, and suppose that for $1 \leqslant i \leqslant n$, $(x_i,y_i)\ (\gamma|\beta_1 \circ \beta_2)\ (x',y')$, say $(x_i,t_i)\ (\gamma|\beta_1)\ (x_i',t_i')$ and $(t_i,y_i)\ (\gamma|\beta_2)\ (t_i',y_i')$. Then $(x_1..x_n\omega,t_1..t_n\omega)\ (\gamma|\beta_1)$ $(x_1'..x_n'\omega,t_1'..t_n'\omega)$ and $(t_1..t_n\omega,y_1..y_n\omega)\ (\gamma|\beta_2)\ (t_1'..t_n'\omega,y_1'..y_n'\omega)$. Thus $(x_1..x_n\omega,y_1..y_n\omega)\ (\gamma|\beta_1 \circ \beta_2)\ (x_1'..x_n'\omega,y_1'..y_n'\omega)$, so $(\gamma|\beta_1 \circ \beta_2) \leqslant {}^2(\beta_1 \circ \beta_2)$. For $(x,y) \in \beta_1 \circ \beta_2$, say $x\ \beta_1\ t\ \beta_2\ y$, $(x,t)\ (\gamma|\beta_1)\ (x,t)$ and $(t,y)\ (\gamma|\beta_2)\ (t,y)$, so $\widehat{(\beta_1 \circ \beta_2)} \leqslant$ $(\gamma|\beta_1 \circ \beta_2) \leqslant {}^2(\beta_1 \circ \beta_2)$. Thus $(\gamma|\beta_1 \circ \beta_2)$ is a congruence on $\beta_1 \circ \beta_2$. Clearly it satisfies condition (C0). It remains by Corollary 224 to check condition (C4). So suppose $(x,x)\ (\gamma|\beta_1 \circ \beta_2)$ (x,y) , say $(x,t)\ (\gamma|\beta_1)\ (x,u)$ and $(t,x)\ (\gamma|\beta_2)\ (u,y)$. Then (C2) for $(\gamma|\beta_1)$ implies $t = u$, and then (C2) for $(\gamma|\beta_2)$ implies $y = z$, as required .]

227 COROLLARY If β_1 and β_2 centralise γ , then $\beta_1 \circ \beta_2$ does.

Proof. By the symmetry of centralising, γ centralises β_1 and β_2 . By Proposition 226, it centralises $\beta_1 \circ \beta_2$, and so again by symmetry $\beta_1 \circ \beta_2$ centralises γ .]

Let α be a congruence on a $\underline{\underline{T}}$-algebra A . Let S be the set of congruences on A centralising α . S is non-empty, since it contains A . Now S is partially ordered by the inclusion relation, and the union of a chain of congruences is the chain's least upper bound, so by Zorn's Lemma S has maximal elements. By Proposition 226, any two such maximal elements are identical. Thus :

228 There is a unique maximal congruence $\eta(\alpha)$ centralising α .

Let γ centralise β on A . By the symmetry of centralising, β centralises γ . By restriction, β and γ centralise $\beta \cap \gamma$. Then by Corollary 227 $\beta \circ \gamma$ centralises $\beta \cap \gamma$. In particular, for any congruence α , $\alpha \circ \eta(\alpha)$ centralises $\alpha \cap \eta(\alpha)$.

The last result of this section enables one to climb to higher levels of centrality :

229 PROPOSITION Let γ centralise β on A .

(i) If $e^i : \beta^{(\gamma|\beta)} \to A^\gamma$; $(x_0,x_1)^{(\gamma|\beta)} \mapsto x_i^\gamma$ for $i = 0, 1$, let $E^i = \ker e^i$. Then E^1 centralises E^0 on $\beta^{(\gamma|\beta)}$.

(ii) For a congruence D on $\beta^{(\gamma|\beta)}$ centralising E^0 , define δ on A by $y \,\delta\, y' \Leftrightarrow \exists\, x, x' \in A .\ (x,y)^{(\gamma|\beta)}\, D\, (x',y')^{(\gamma|\beta)}$. Then δ is a congruence on A centralising β .

(iii) If $D > E^1$, then $\delta > \gamma$. Thus if $\gamma = \eta(\beta)$, $E^1 = \eta(E^0)$.

<u>Proof.</u> (i) If $(x,y)^{(\gamma|\beta)}\, E^0\, (x',y')^{(\gamma|\beta)}$, $x \,\gamma\, x'$, so by (C1) $\exists\, y'' \in A .\ (x',y')\, (\gamma|\beta)\, (x,y'')$. Thus $((x,y)^{(\gamma|\beta)},(x',y')^{(\gamma|\beta)})$ in E^0 may be written as $((x,y)^{(\gamma|\beta)},(x,y'')^{(\gamma|\beta)})$. Now define $(E^1|E^0)$ on E^0 by $((x,y)^{(\gamma|\beta)},(x,z)^{(\gamma|\beta)})\ (E^1|E^0)$ $((x',y')^{(\gamma|\beta)},(x',z')^{(\gamma|\beta)}) \Leftrightarrow (y,z)\ (\gamma|\beta)\ (y',z')$. Then $(E^1|E^0)$ is a congruence on E^0 , and by Corollary 224 it centres.

(ii) Clearly, $\delta \leqslant {}^2A$. Let $y \in A$. Then $((y,y)^{(\gamma|\beta)},(y,y)^{(\gamma|\beta)}) \in E^0$, and by reflexivity of $(D|E^0)$, $((y,y)^{(\gamma|\beta)},(y,y)^{(\gamma|\beta)})\ (D|E^0)\ ((y,y)^{(\gamma|\beta)},(y,y)^{(\gamma|\beta)})$. Thus by (C0), $(y,y)^{(\gamma|\beta)}\, D\, (y,y)^{(\gamma|\beta)}$. Hence $y \,\delta\, y$, and so δ is a congruence on A . Define $(\delta|\beta)$ on β by $(y,x)\ (\delta|\beta)\ (y',x') \Leftrightarrow$ $((x,y)^{(\gamma|\beta)},(x,x)^{(\gamma|\beta)})\ (D|E^0)\ ((x',y')^{(\gamma|\beta)},(x',x')^{(\gamma|\beta)})$. Clearly $(\delta|\beta) \leqslant {}^2\beta$. Let $(y,x) \in \beta$. Then $(x,y)^{(\gamma|\beta)}\, E^0\, (x,x)^{(\gamma|\beta)}$, so by reflexivity of $(D|E^0)$, $((x,y)^{(\gamma|\beta)},(x,x)^{(\gamma|\beta)})$ $(D|E^0)$ $((x,y)^{(\gamma|\beta)},(x,x)^{(\gamma|\beta)})$. Thus (y,x) $(\delta|\beta)$ (y,x) , i.e. $\hat{\beta} \leqslant (\delta|\beta)$, and $(\delta|\beta)$ becomes a congruence on β . By (C0) for $(D|E^0)$, (C0) for $(\delta|\beta)$ is satisfied. Suppose (x,x) $(\delta|\beta)$ (x,y) . Then $((x,x)^{(\gamma|\beta)},(x,x)^{(\gamma|\beta)})$ $(D|E^0)$ $((y,x)^{(\gamma|\beta)},(y,y)^{(\gamma|\beta)})$. By (C3) for $(D|E^0)$, (y,x) $(\gamma|\beta)$ (y,y) . Then by (C4) for $(\gamma|\beta)$, $x = y$. Thus (C4) for $(\delta|\beta)$ is satisfied. By Corollary 224, δ centralises β by means of $(\delta|\beta)$.

(iii) Let $((x,y)^{(\gamma|\beta)},(x',y')^{(\gamma|\beta)}) \in D - E^1$. Then $(y,y') \in \delta - \gamma$.]

2.3 Nilpotence.

This section uses the ideas of centrality introduced earlier in this chapter to define nilpotence for Mal'cev algebras. To begin with, some elementary properties of the operator η are needed. Let A be in a Mal'cev variety $\underline{\underline{T}}$.

231 PROPOSITION Let α be a congruence on A .

(i) If $B \leqslant A$, then $\eta(\alpha) \cap {}^2B \leqslant \eta(\alpha \cap {}^2B)$.

(ii) If $\theta : A \to A\theta$ is an epimorphism, then $\eta(\alpha)^2\theta \leqslant \eta(\alpha^2\theta)$.

<u>Proof.</u> (i) $(\alpha \cap {}^2B | \eta(\alpha) \cap {}^2B) = (\alpha|\eta(\alpha)) \cap {}^2(\eta(\alpha) \cap {}^2B)$.

(ii) $(\alpha^2\theta | \eta(\alpha)^2\theta) = (\alpha|\eta(\alpha))^4\theta$.]

Let β and γ be congruences on A . Let $K(\beta,\gamma) = \{ \hat{A} \leqslant \delta \leqslant {}^2A \mid (\beta \circ \delta)^2 \mathrm{nat}\delta < \eta(\gamma^2 \mathrm{nat}\delta) \}$. Since $\beta^2\mathrm{nat}\beta = \widehat{A\mathrm{nat}\beta} \leqslant \eta(\gamma^2\mathrm{nat}\beta)$, $\beta \in K(\beta,\gamma)$. Define the <u>commutator</u> of β and γ to be $[\beta,\gamma] = \bigcap_{\delta \in K(\beta,\gamma)} \delta$. Note that $[\beta,\gamma] \leqslant \beta$. Define $(\beta^2\mathrm{nat}[\beta,\gamma] | \gamma^2\mathrm{nat}[\beta,\gamma])$ by $(x^{[\beta,\gamma]}, y^{[\beta,\gamma]})$ $(\beta^2\mathrm{nat}[\beta,\gamma] | \gamma^2\mathrm{nat}[\beta,\gamma])$ $(z^{[\beta,\gamma]}, t^{[\beta,\gamma]}) \Leftrightarrow \forall\ \delta \in K(\beta,\gamma)$, (x^δ, y^δ) $((\beta \circ \delta)^2\mathrm{nat}\delta | \gamma^2\mathrm{nat}\delta)$ (z^δ, t^δ) . Then

232 $\beta^2\mathrm{nat}[\beta,\gamma] \leqslant \eta(\gamma^2\mathrm{nat}[\beta,\gamma])$, i.e. $[\beta,\gamma] \in K(\beta,\gamma)$.

If α is a congruence on A centralised by ${}^2A$, α is said to be <u>central</u>. The unique maximal central congruence $\eta({}^2A)$ is called <u>the centre congruence</u> $\zeta(A)$ (or just ζ) . A <u>central series</u> in A is a series $\hat{A} = \alpha_0 \leqslant \alpha_1 \leqslant \ldots \leqslant \alpha_n = {}^2A$ of congruences on A such that for $i = 1 , \ldots , n$, $\alpha_i{}^2\mathrm{nat}\alpha_{i-1} \leqslant \zeta(A\mathrm{nat}\alpha_{i-1})$. Writing $A_0 = {}^2A$, $\zeta_0(A) = \hat{A}$, define $A_{i+1} = [A_i, A_0]$ and $\zeta_{i+1}(A)$ (or just ζ_{i+1}) by $\zeta_{i+1}{}^2\mathrm{nat}\zeta_i = \zeta(A\mathrm{nat}\zeta_i)$ inductively. Note that $A_0 \geqslant A_1 \geqslant \ldots$ and $\zeta_0 \leqslant \zeta_1 \leqslant \ldots$. If $A_j \leqslant \zeta_i$, then $(A_j \circ \zeta_{i-1})^2\mathrm{nat}\zeta_{i-1} \leqslant \zeta_i{}^2\mathrm{nat}\zeta_{i-1} = \zeta(A\mathrm{nat}\zeta_{i-1})$, so $\zeta_{i-1} \geqslant [A_j, A_0]$

$= A_{j+1}$. Also $(A_{j-1} \circ \zeta_i)^2 \mathrm{nat}\zeta_i \leqslant \zeta(A\mathrm{nat}\zeta_i) = \zeta_{i+1}\mathrm{nat}\zeta_i$, so $A_{j-1} \leqslant \zeta_{i+1}$. Thus $A_c \leqslant \zeta_0$ iff $A_0 \leqslant \zeta_c$. If A satisfies these conditions, c being the least integer for which it does, then A is said to be <u>nilpotent of class</u> c . $A_0 \geqslant A_1 \geqslant \ldots \geqslant A_c$ is called the <u>lower central series</u> and $\zeta_0 \leqslant \zeta_1 \leqslant \ldots \leqslant \zeta_c$ the <u>upper central series</u>. The variety of all $\underline{\underline{T}}$-algebras of class at most c is denoted by $\underline{\underline{N}}_c(\underline{\underline{T}})$ or just $\underline{\underline{N}}_c$. The class of all nilpotent $\underline{\underline{T}}$-algebras, of unspecified nilpotency class, is denoted by $\underline{\underline{N}}$. The symbol $\underline{\underline{Z}}$ will be used instead of $\underline{\underline{N}}_1$. For example, if $\underline{\underline{T}}$ is the variety of groups, loops, Lie algebras, then $\underline{\underline{Z}}(\underline{\underline{T}})$ is the variety of abelian groups, abelian groups, abelian Lie algebras respectively.

233 If α on A is central, $^2(\alpha\mathrm{nat}(^2A|\alpha)) = \ker(e^i : \alpha\mathrm{nat}(^2A|\alpha) \to A\mathrm{nat}^2A)$, so by Proposition 229(i) $\alpha\mathrm{nat}(^2A|\alpha) \in \underline{\underline{Z}}$. $\alpha\mathrm{nat}(^2A|\alpha)$ is rather special, as by (C3) of Proposition 223 it has the singleton subalgebra $\{\hat{A}\}$. Its properties will be exploited in Chapter 4.

A subalgebra B of A is said to be <u>normal</u>, written $B \lhd A$, if B is an equivalence class of $< \hat{A} , {}^2B >$, and then $A\mathrm{nat}< \hat{A} , {}^2B >$ is denoted by A/B . Note that by Proposition 223, a congruence α on A is central iff $\hat{A} \lhd \alpha$ (cf. [Hu, Satz I.9.14]).

234 PROPOSITION A variety $\underline{\underline{Y}}$ of $\underline{\underline{T}}$-algebras is a subvariety of $\underline{\underline{Z}}(\underline{\underline{T}})$ iff every subalgebra B of every $\underline{\underline{Y}}$-algebra A is normal in A .

Proof. Suppose the latter condition is satisfied. If A is a $\underline{\underline{Y}}$-algebra, so is ${}^2A$. Then the subalgebra $\hat{A}$ of ${}^2A$ is normal. Proposition 223 then implies that ${}^2A$ is a central congruence, so that $A \in \underline{\underline{Z}}(\underline{\underline{T}})$.

Conversely, if B is a subalgebra of a $\underline{\underline{Z}}(\underline{\underline{T}})$-algebra A , then

$$\begin{aligned} < \hat{A} , {}^2B > &= ({}^2B)\mathrm{nat}({}^2A|{}^2A) \\ &= \{ (a,a') \in {}^2A \mid \exists (b,b') \in {}^2B . \\ &\qquad (b,b') ({}^2A|{}^2A) (a,a') \} . \end{aligned}$$

If $(b,x) \in < \hat{A} , {}^2B >$, then $\exists (b',b'') \in {}^2B . (b,x) ({}^2A|{}^2A) (b',b'')$. Then $x = (b,b',b'')P \in B$. Thus B is an equivalence class of $< \hat{A} , {}^2B >$, i.e. $B \lhd A$.]

This section concludes with one application of nilpotence, to the study of non-generators and the Frattini subalgebra of the Mal'cev algebra A . Until the end of the current chapter, the term "algebra" will no longer exclude the empty algebra if the variety $\underline{\underline{T}}$ has no nullary operations. An element x of A is called a <u>non-generator</u> of A if $< x , S > = A$ for any subset S of A implies $< S > = A$. Note that if $\underline{\underline{T}}$ has no nullary operations,

$< \emptyset > = \emptyset$. If $\underline{\underline{T}}$ has the nullary operation η , then η regarded as an element of A is a non-generator, since $\eta \in < S >$ for any subset S of A . It may happen that A has no non-generators. For example, let $\underline{\underline{T}}$ be the variety of quasigroups, and let $A = \{1,2,3\}$ with multiplication table

.	1	2	3
1	1	3	2
2	3	2	1
3	2	1	3

.

Every pair of elements of A generates A , but no singleton does. Thus A has no non-generators. Note too that the intersection of all the maximal subquasigroups $\{1\}$, $\{2\}$, $\{3\}$ is the empty quasigroup.

As in group theory, one has the

235 PROPOSITION The set $\varphi(A)$ of non-generators of A forms a subalgebra called the <u>Frattini subalgebra</u>. It is the intersection of all the maximal subalgebras of A .

<u>Proof.</u> Let ω be an n-ary operation on A , let $S \subseteq A$, and let $x_1 ,\ldots, x_n \in \varphi(A)$. Then $< x_1 \ldots x_n\omega , S > = A \Rightarrow < x_1,\ldots,x_n,S > = A \Rightarrow < x_2,\ldots,x_n,S > = A \Rightarrow \ldots \Rightarrow < x_n,S > = A \Rightarrow < S > = A$. Thus $\varphi(A)$ is a subalgebra of A .

Let $x \in A$. If there is a maximal subalgebra M not containing x , then $< x,M > = A$ but $< M > = M < A$. Thus x is not a non-generator. So each non-generator is contained in the intersection of all the maximal subalgebras.

Conversely, suppose the element x of A is contained in all maximal subalgebras. Let $< x,S > = A$ for $S \subseteq A$. If $< S > < A$, $x \notin S$. Using Zorn's Lemma, choose a subalgebra $M \geqslant < S >$ maximal with respect to the requirement $x \notin M$. Now $< x,M > \geqslant < x,S > = A$; by choice of M any subalgebra of A properly containing M must contain x as well, and so all of A . Thus M is a maximal subalgebra not containing x . But x is in all maximal subalgebras. So $< S > = A$, and x is a non-generator.]

In groups, $\varphi(A)$ is a characteristic subalgebra of A , and so is normal in A . For other algebras characteristic subalgebras may not necessarily be normal, and so Frattini subalgebras need not be. But the Frattini subalgebra of a nilpotent Mal'cev algebra, if non-empty, is normal. The connection with centrality comes from the following

236 PROPOSITION If M is a subalgebra of A , and $M\mathrm{nat}\zeta(A) = A$

(i.e. every element of A is in the ζ-class of some element of M), then $M \lhd A$ and $A/M \in \underline{\underline{Z}}(\underline{\underline{T}})$.

<u>Proof.</u> $M\mathrm{nat}\zeta = A$, so if $a \in A$, $\exists\, m \in M$. $(m,a) \in \zeta$. Define a congruence μ on A by $(a,a') \in \mu \Leftrightarrow \exists\, m , m' \in M$. $(a,m)\ ({}^2A|\zeta)\ (a',m')$. If $m , m' \in M$, then $(m,m)\ ({}^2A|\zeta)\ (m',m')$ by (RR) for $({}^2A|\zeta)$, so $m\, \mu\, m'$. Conversely, if $x\, \mu\, m \in M$, say $(x,m'')\ ({}^2A|\zeta)\ (m,m')$, then $x = (m,m',m'')P \in M$. Thus M is an equivalence class of μ . But $(a,a') \in \mu \Rightarrow a' = (a,m,m')P \Rightarrow (a,a') = ((a,a),(m,m),(m,m'))P \in < \hat{A} , {}^2M >$. Thus $M \lhd A$. Since $M\mathrm{nat}\zeta = A$, $\mu \circ \zeta = {}^2A$. Thus $\zeta^2\mathrm{nat}\mu = {}^2(A/M)$. But by Proposition 231(ii), $\zeta^2\mathrm{nat}\mu \leqslant \zeta(A/M)$. Thus $A/M \in \underline{\underline{Z}}(\underline{\underline{T}})$.]

Applying this, one gets the

237 THEOREM If A is nilpotent and has non-generators, then $\varphi(A) \lhd A$ and $A/\varphi(A) \in \underline{\underline{Z}}(\underline{\underline{T}})$.

<u>Proof.</u> Suppose $A \in \underline{\underline{N}}_c$. Then $\zeta_c = {}^2A$. Let M be a maximal subalgebra of A . Then $M\mathrm{nat}\zeta_c(A) = A$. Let n be minimal with respect to the requirement $M\mathrm{nat}\zeta_n = A$. Note that $n > 0$, since $M\mathrm{nat}\zeta_0 = M$. Also $i < n \Rightarrow M\mathrm{nat}\zeta_i \leqslant M$, since $M\mathrm{nat}\zeta_i < A$. Now $(M\mathrm{nat}\zeta_{n-1})\mathrm{nat}(\zeta_n{}^2\mathrm{nat}\zeta_{n-1}) = A\mathrm{nat}\zeta_{n-1}$. Thus by Proposition 236, $M\mathrm{nat}\zeta_{n-1} \lhd A\mathrm{nat}\zeta_{n-1}$ and $(A\mathrm{nat}\zeta_{n-1})/(M\mathrm{nat}\zeta_{n-1}) \in \underline{\underline{Z}}$. Hence $M \lhd A$

and $A/M \in \underline{\underline{Z}}(\underline{\underline{T}})$. Now by Proposition 235, $\varphi(A)$ is the intersection of all such M, and by hypothesis $\varphi(A)$ is non-empty, so $\varphi(A) \lhd A$ and $A/\varphi(A) \in \underline{\underline{Z}}$.]

3 Direct decompositions

Let $\underline{\underline{T}}$ be a Mal'cev variety, and let α be a congruence on the $\underline{\underline{T}}$-algebra A. Ore's Theorem 132 states that if α has both the maximal and minimal conditions on subcongruences, then any factorisation $\alpha = \alpha_1 \sqcap \cdots \sqcap \alpha_m$ as a direct product of indecomposables is unique up to isomorphism of congruences. The first section of this chapter proves this theorem and related results by a Fitting's Lemma method analogous to the usual group-theoretic proof of the Krull-Remak-Schmidt Theorem [Hu, Satz I.12.3]. The results of that section are described as "classical", because they do not involve centrality. Then in the second section centrality is brought in to weaken the hypothesis of the unique factorisation theorem to the maximal and minimal conditions for $\alpha \wedge \eta(\alpha)$, and to strengthen the conclusion to uniqueness up to "central isomorphism of congruences with respect to α ". This is probably the best possible result for Mal'cev algebras, since it corresponds to the best possible result for groups.

Throughout this chapter, Greek letters except φ and σ denote congruences on the $\underline{\underline{T}}$-algebra A .

3.1 Classical results.

The first result is a form of <u>Fitting's Lemma</u> in the category of congruences :

311 PROPOSITION If $\varphi : \beta \to \beta$ is an endomorphism of congruences, and β has both the maximal and minimal conditions on subcongruences, then there is a positive integer n such that $\beta = \beta\varphi^n \cap \mathrm{Ker}\varphi^n$.

<u>Proof.</u> $\beta \geqslant \beta\varphi \geqslant \beta\varphi^2 \geqslant \ldots$ and $\mathrm{Ker}\varphi \leqslant \mathrm{Ker}\varphi^2 \leqslant \mathrm{Ker}\varphi^3 \leqslant \ldots$.
The maximal and minimal conditions yield the existence of a positive integer n such that $\beta\varphi^n = \beta\varphi^{n+1} = \ldots$ and $\mathrm{Ker}\varphi^n = \mathrm{Ker}\varphi^{n+1} = \ldots$.
Suppose $(x,y) \in \beta\varphi^n \cap \mathrm{Ker}\varphi^n$. Then $(x,y)\varphi^n = (x,x)$ and also $\exists\, (x,z) \in \beta \,.\; (x,z)\varphi^n = (x,y)$. $\Rightarrow (x,z)\varphi^{2n} = (x,y)\varphi^n = (x,x)$. $\Rightarrow (x,z) \in \mathrm{Ker}\varphi^{2n} = \mathrm{Ker}\varphi^n$. $\Rightarrow (x,z)\varphi^n = (x,y) = (x,x)$. Hence $\beta\varphi^n \cap \mathrm{Ker}\varphi^n = \hat{A}$.

Let $(x,y) \in \beta$. Let $(x,z) = (x,y)\varphi^n \in \beta\varphi^n = \beta\varphi^{2n}$.

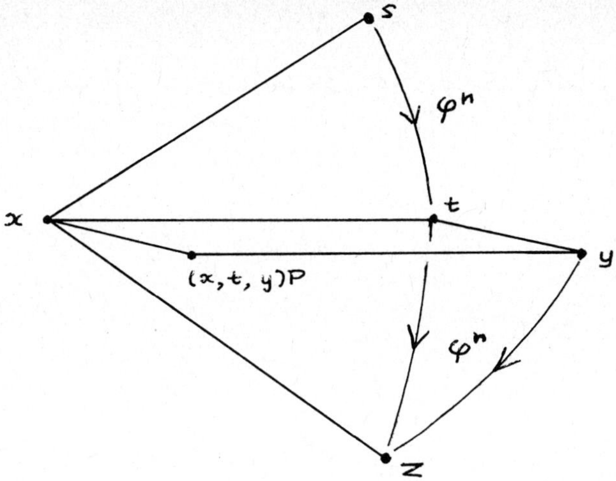

Then $\exists$ $(x,s) \in \beta$. $(x,z) = (x,s)\varphi^{2n}$. Let $(x,s)\varphi^n = (x,t)$.

Then
$$(x,x)\varphi^n = (x,x) \quad ,$$
$$(x,t)\varphi^n = (x,z) \quad ,$$
$$\text{and} \quad (x,y)\varphi^n = (x,z) \quad .$$

$\Rightarrow$ $(x,(x,t,y)P)\varphi^n = (x,x)$. Thus $(x,(x,t,y)P) \in \mathrm{Ker}\varphi^n$.

Also
$$(x,s)\varphi^n = (x,t) \quad ,$$
$$(t,t)\varphi^n = (t,t) \quad ,$$
$$\text{and} \quad (y,y)\varphi^n = (y,y) \quad .$$

$\Rightarrow$ $((x,t,y)P,(s,t,y)P)\varphi^n = ((x,t,y)P,y)$. Thus $((x,t,y)P,y) \in \beta\varphi^n$.

So $x \ \mathrm{Ker}\varphi^n \ (x,t,y)P \ \beta\varphi^n \ y$. Hence $\beta = \mathrm{Ker}\varphi^n \circ \beta\varphi^n$.]

312 COROLLARY If $\beta\varphi = \beta$, then φ is an automorphism of congruences.

<u>Proof.</u> $\beta = \beta\varphi \Rightarrow \beta = \beta\varphi^n$ for all positive integers n . Thus by

Proposition 3.11, there is an integer n with $\mathrm{Ker}\varphi^n = \hat{A}$. But $\mathrm{Ker}\varphi \leqslant \mathrm{Ker}\varphi^n$, and so $\mathrm{Ker}\varphi = \hat{A}$. Thus φ injects. Clearly it also surjects.]

Now let β be an indecomposable congruence, and let $\beta \cap \gamma = \alpha_0 \cap \alpha_1$. For $i = 0, 1$, define $\varphi_i : \beta \to \beta$; $(x,y) \mapsto (x,y_2)$ where $x\ \alpha_i\ y\ \alpha_{1-i}\ y$ and $x\ \beta\ y_2\ \gamma\ y_1$:

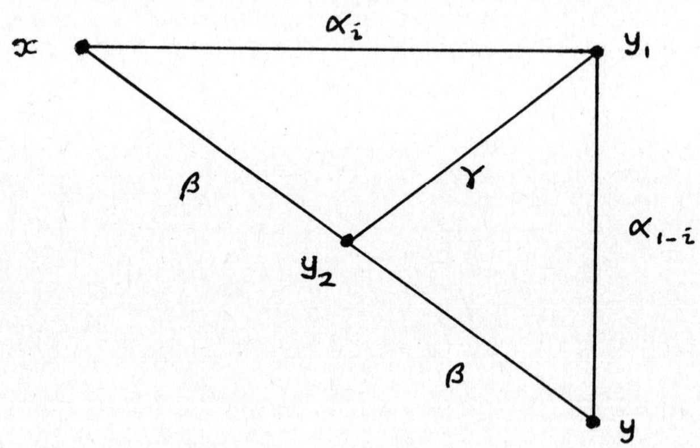

φ_0 and φ_1 commute, for one has the following diagram in which all the vertices are specified uniquely by (x,y) and the relations shown :

313 :

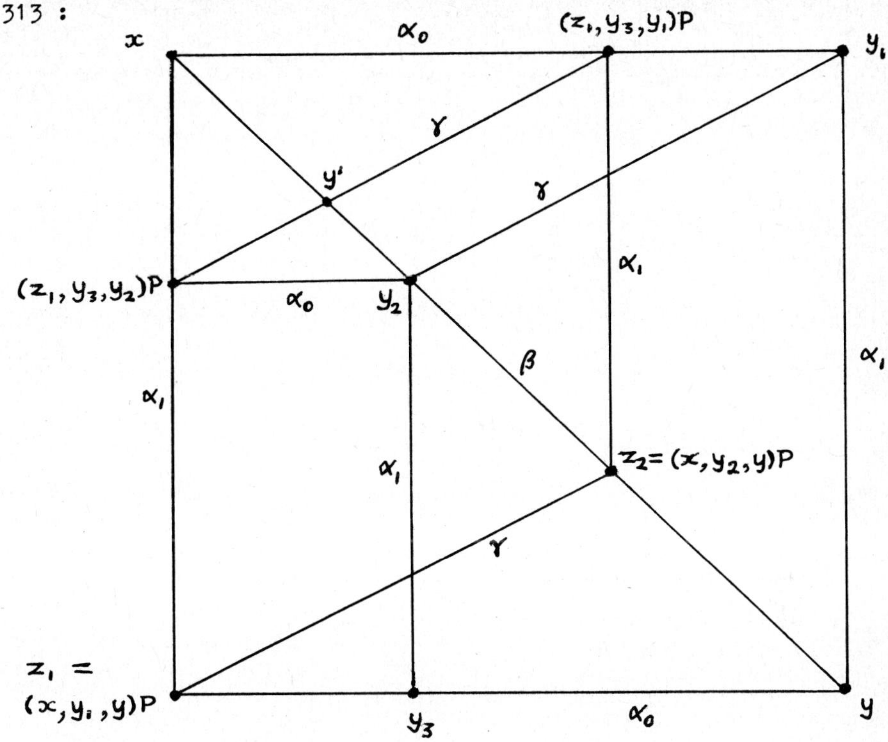

so that $(x,y)\varphi_0\varphi_1 = (x,y_2)\varphi_1 = (x,y') = (x,z_2)\varphi_0 = (x,y)\varphi_1\varphi_0$.

By Proposition 311, there is a positive integer n such that $\beta = \beta\varphi_0^n \cap \mathrm{Ker}\varphi_0^n = \beta\varphi_1^n \cap \mathrm{Ker}\varphi_1^n$. Since β is indecomposable, one of each pair of factors is just $\hat{A}$. Suppose that neither φ_0 nor φ_1 is an automorphism of congruences. Then by Corollary 312 $\beta\varphi_i < \beta$ for each i , so that $\beta\varphi_i^n = \hat{A}$. Now in the notation of 313, $y = (y_2,x,z_2)P$, since $(y_2,x,z_2)P \;\beta\; y = (y_1,x,z_1)P \;\gamma\; (y_2,x,z_2)P$ and $\beta \wedge \gamma = \hat{A}$. In other words

$(x,y) = ((x,y)\varphi_0,(x,x),(x,y)\varphi_1)P$. Thus $\beta \leqslant \langle \beta\varphi_0, \beta\varphi_1 \rangle$. Suppose as an induction hypothesis on m that $\beta \leqslant \langle \beta\varphi_0^r\varphi_1^{m-r} \mid r = 0,1,..,m \rangle$. Then $\beta\varphi_0^r\varphi_1^{m-r} \leqslant \langle \beta\varphi_0^{r+1}\varphi_1^{m-r} , \beta\varphi_0^r\varphi_1^{m-r+1} \rangle$. Thus $\beta \leqslant \langle \beta\varphi_0^r\varphi_1^{(m+1)-r} \mid r = 0,1,..,(m+1) \rangle$. The result is true for all m , and in particular for $m = 2n$. Hence $\beta = \hat{A}$. But β is indecomposable, so this is impossible. Thus $\exists\, i \in \{0,1\}$. φ_i is an automorphism of congruences.

Suppose that φ_i is an automorphism of congruences. Since it is clear from the context that projections onto α_i must be along α_{1-i} , the suffix in the projection notation will be dropped. Now $\alpha_i^\beta \cap \gamma = \hat{A}$, for if $x\ \alpha_i^\beta \cap \gamma\ y$, say with $x\ \beta\ z\ \alpha_{1-i}\ y\ \alpha_i\ x$, then $(x,z)\varphi_i = (x,x)$, so that $z = x = y$. As $\beta = \beta\varphi_i \leqslant \alpha_i^\beta \sqcap \gamma$, it follows that $\alpha_i^\beta \sqcap \gamma = \beta \sqcap \gamma$. By the modular law in the lattice of congruences, $\alpha_i = (\alpha_i^\beta \sqcap \gamma) \cap \alpha_i = \alpha_i^\beta \sqcap (\gamma \cap \alpha_i)$. If $(x,y) \in \beta \cap ((\gamma \cap \alpha_i) \sqcap \alpha_{1-i})$, then $(x,y)\varphi_i = (x,x)$, so that $y = x$:
Thus $\beta \cap ((\gamma \cap \alpha_i) \sqcap \alpha_{1-i}) = \hat{A}$.

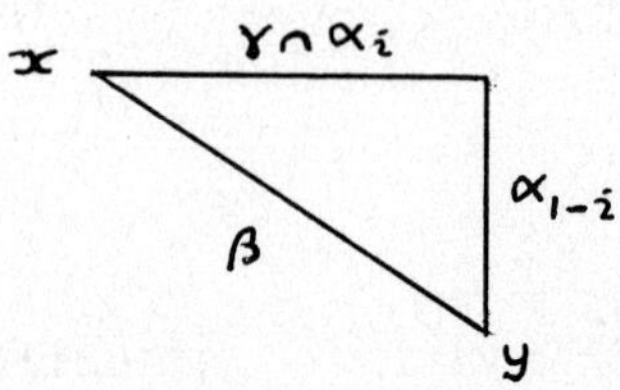

Now $\beta \leqslant \alpha_i^\beta \sqcap \alpha_{1-i}^\beta$, and so $\beta \sqcap \alpha_{1-i}^\beta \leqslant \alpha_i^\beta \sqcap \alpha_{1-i}^\beta$. Suppose $(x,y) \in \alpha_i^\beta \sqcap \alpha_{1-i}^\beta$, say $\exists\, s$. $x\ \alpha_i^\beta\ s\ \alpha_{1-i}^\beta\ y$. Since $(x,s) \in \alpha_i^\beta$, $\exists\, t$. $x\ \beta\ t\ \alpha_{1-i}^\beta\ s$:

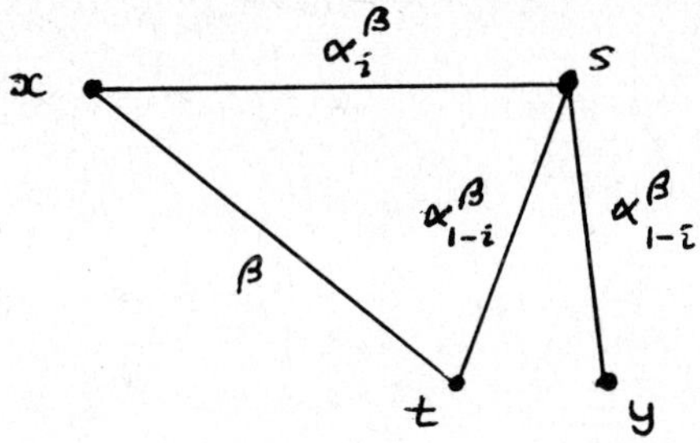

Then $y\ \alpha_{1-i}^\beta\ t\ \beta\ x$, i.e. $(x,y) \in \beta \sqcap \alpha_{1-i}^\beta$. Thus $\beta \sqcap \alpha_{1-i}^\beta =$ $\alpha_i^\beta \sqcap \alpha_{1-i}^\beta$. $\Rightarrow \alpha_i^\beta \leqslant \beta \sqcap \alpha_{1-i}^\beta \leqslant \beta \sqcap (\gamma \wedge \alpha_i) \sqcap \alpha_{1-i}$. $\Rightarrow \alpha_0 \sqcap \alpha_1 =$ $\alpha_i^\beta \sqcap (\gamma \wedge \alpha_i) \sqcap \alpha_{1-i} \leqslant \beta \sqcap (\gamma \wedge \alpha_i) \sqcap \alpha_{1-i}$. Summarising,

314 PROPOSITION Let β be an indecomposable congruence, and let $\beta \sqcap \gamma = \alpha_0 \sqcap \alpha_1$. Define $\varphi_i : \beta \to \beta\ ;\ (x,y) \mapsto (x,y'')$, where $x\ \alpha_i\ y'\ \alpha_{1-i}\ y$ and $y'\ \gamma\ y''\ \beta\ x$. Then $\exists\ j \in \{0,1\}$. φ_j is an automorphism of congruences. If φ_i is an automorphism of congruences, $\alpha_i = \alpha_i^\beta \sqcap (\gamma \wedge \alpha_i)$ and $\beta \sqcap \gamma = \alpha_i^\beta \sqcap \gamma =$ $\beta \sqcap (\gamma \wedge \alpha_i) \sqcap \alpha_{1-i} = \alpha_0 \sqcap \alpha_1$.]

By induction, Proposition 314 can be used to prove the following, known as the Classical Exchange Theorem :

315 THEOREM If indecomposable β has the maximal and minimal conditions on subcongruences, and $\beta \sqcap \gamma = \delta_1 \sqcap \delta_2 \sqcap \cdots \sqcap \delta_n$, then

$\exists\ 1 \leqslant l \leqslant n\ .\ \delta_l = \xi \sqcap \varepsilon$ and $\beta \sqcap \gamma = \xi \sqcap \gamma =$

$$\beta \sqcap \varepsilon \sqcap \delta_1 \sqcap \cdots \sqcap \delta_{l-1} \sqcap \delta_{l+1} \sqcap \cdots \sqcap \delta_n .$$

<u>Proof.</u> This follows by induction on n. If $n = 2$, it is just Proposition 314. Suppose it holds for all decompositions of $\beta \sqcap \gamma$ with less than n factors. Let $\alpha_0 = \delta_2 \sqcap \cdots \sqcap \delta_n$, $\alpha_1 = \delta_1$. By Proposition 314, either φ_0 or φ_1 is an automorphism of congruences. If φ_1 is an automorphism, the result follows with $l = 1$ from Proposition 314. Otherwise, Proposition 314 gives that $\beta \sqcap \gamma = \alpha_0^\beta \sqcap \gamma = \beta \sqcap (\gamma \cap \alpha_0) \sqcap \alpha_1 = \alpha_0 \sqcap \alpha_1$. In particular, β is isomorphic with α_0^β, so α_0^β is indecomposable. Proposition 314 also gives that $\alpha_0^\beta \sqcap (\gamma \cap \alpha_0) = \delta_2 \sqcap \cdots \sqcap \delta_n$, so by the induction hypothesis $\exists\ 2 \leqslant l \leqslant n\ .\ \delta_l = \xi \sqcap \varepsilon$ and $\alpha_0^\beta \sqcap (\gamma \cap \alpha_0) = \xi \sqcap (\gamma \cap \alpha_0) = \alpha_0^\beta \sqcap \varepsilon \sqcap \delta_2 \sqcap \cdots \sqcap \delta_{l-1} \sqcap \delta_{l+1} \sqcap \cdots \sqcap \delta_n = \delta_2 \sqcap \cdots \sqcap \delta_n$.

Define $f : \beta \to \alpha_0^\beta\ ;\ (x,y) \mapsto (x,y')$ where $x\ \alpha_0^\beta\ y'\ \delta_1\ y$, recalling that $\delta_1 = \alpha_1$ and $\beta \sqcap \gamma = \alpha_0 \sqcap \alpha_1$. Define $g : \alpha_0^\beta \to \beta\ ;\ (x,y) \mapsto (x,z)$, where $x\ \beta\ z\ \delta_1\ y$. This is well-defined, since $\beta \sqcap \gamma = \beta \sqcap (\gamma \cap \alpha_0) \sqcap \alpha_1$ implies $\beta \cap \delta_1 = \hat{A}$. f and g are mutually inverse isomorphisms of congruences. Define $f' : \alpha_0^\beta \to \delta_l\ ;\ (x,y) \mapsto (x,y'')$, where $x\ \delta_l\ y''\ \delta_2 \sqcap \cdots \sqcap \delta_{l-1} \sqcap \delta_{l+1} \sqcap \cdots \sqcap \delta_n\ y$. Define $g' : \delta_l \to \alpha_0^\beta\ ;\ (x,y) \mapsto (x,z')$, where $x\ \alpha_0^\beta\ z'\ \delta_2 \sqcap \cdots \sqcap \delta_{l-1} \sqcap \delta_{l+1} \sqcap \cdots \sqcap \delta_n\ y$. f' and g' are

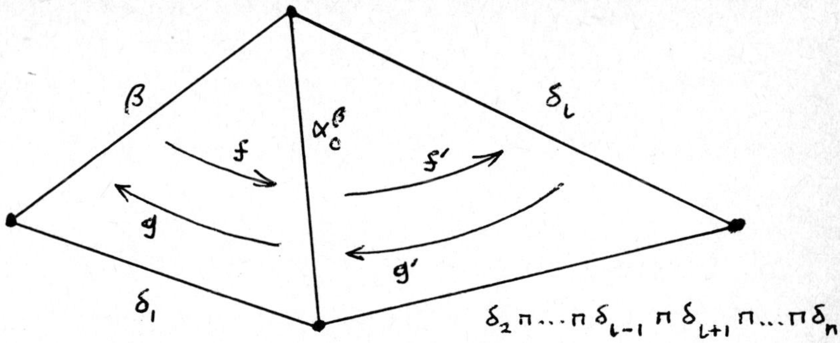

mutually inverse isomorphisms of congruences. Thus $ff'g'g : \beta \to \beta$ is an automorphism of congruences. Setting $\alpha_0 = \delta_l$, $\alpha_1 = \delta_1 \sqcap \cdots \sqcap \delta_{l-1} \sqcap \delta_{l+1} \sqcap \cdots \sqcap \delta_n$ in Proposition 314, this means that φ_0 is an automorphism of congruences. The result follows from the second statement of Proposition 314.]

Now for the <u>Classical Unique Factorisation Theorem</u> :

316 THEOREM Let α be a congruence on the $\underline{\underline{T}}$-algebra A . If α has both the maximal and minimal conditions on subcongruences, then it has a unique factorisation as a direct product of indecomposables. More precisely, if $\alpha = \alpha_1 \sqcap \alpha_2 \sqcap \cdots \sqcap \alpha_m = \beta_1 \sqcap \beta_2 \sqcap \cdots \sqcap \beta_n$ with the α_i and β_j indecomposable, then $m = n$ and there is a permutation σ of $\{1,\ldots,m\}$ such that

317 : $\alpha = \beta_{\sigma(1)} \sqcap \cdots \sqcap \beta_{\sigma(r)} \sqcap \alpha_{r+1} \sqcap \cdots \sqcap \alpha_m$

for $r = 0,\dots,m$. Further, α_i is isomorphic with $\beta_{\sigma(i)}$ for $i = 1,\dots,m$.

Proof. Since α has the minimal condition on subcongruences, such decompositions exist by Proposition 131. Suppose $\sigma : \{1,\dots,s\} \to \{1,\dots,n\}$ has been defined so that 317 holds for $r \leqslant s$. In particular, $\alpha_{s+1} \sqcap (\beta_{\sigma(1)} \sqcap \cdots \sqcap \beta_{\sigma(s)} \sqcap \alpha_{s+2} \sqcap \cdots \sqcap \alpha_m) = \beta_1 \sqcap \cdots \sqcap \beta_n$. By the Classical Exchange Theorem 315, $\exists\ 1 \leqslant l \leqslant n$. $\beta_l = \xi \sqcap \varepsilon$ and $\alpha_{s+1} \sqcap (\beta_{\sigma(1)} \sqcap \cdots \sqcap \beta_{\sigma(s)} \sqcap \alpha_{s+2} \sqcap \cdots \sqcap \alpha_m)$

$$= \xi \sqcap (\beta_{\sigma(1)} \sqcap \cdots \sqcap \beta_{\sigma(s)} \sqcap \alpha_{s+2} \sqcap \cdots \sqcap \alpha_m) .$$

Now α_{s+1} is isomorphic to ξ , and so $\xi > \hat{A}$. Thus $\varepsilon = \hat{A}$, $\beta_l = \xi$. Note that $\xi \sqcap (\beta_{\sigma(1)} \sqcap \cdots \sqcap \beta_{\sigma(s)}) = \hat{A}$, so $l \notin \{\sigma(1),\dots,\sigma(s)\}$. Define $\sigma(s+1) = l$. Then $\alpha = \beta_{\sigma(1)} \sqcap \cdots \sqcap \beta_{\sigma(s+1)} \sqcap \alpha_{s+2} \sqcap \cdots \sqcap \alpha_m$. 317 follows for all r by induction. The isomorphism between α_i and $\beta_{\sigma(i)}$ comes from comparing 317 for $r = i-1$ and $r = i$.]

318 REMARK Apart from its use in proving the Classical Unique Factorisation Theorem 316, Fitting's Lemma 311 is of independent interest. An example of this is furnished by entropic quasigroups, introduced and studied by Murdoch [Mu] under the now confusing name of "abelian quasigroups" (abelian groups are also quasigroups, after all, and there is also Manin's notion of "Abelian quasigroup" as in Section 4.3). Entropic quasigroups are, in the language of Section

1.2, quasigroups in the variety of quasigroups : they satisfy the <u>entropic law</u> $(a.b).(c.d) = (a.c).(b.d)$. Murdoch gave three structure theorems for entropic quasigroups in [Mu]. His First Structure Theorem decomposed a finite entropic quasigroup Q into a direct product of two quasigroups. Define a quasigroup morphism $r : Q \to Q ; x \mapsto x^r$ where $x.x^r = x$. The element x^r of Q is called the <u>right unit</u> of x . Murdoch's decomposition produced one factor in which every element was a right unit and a direct complement of that factor having a unique idempotent. His proof considered extensions of the limit of the series

$$Q \to Qr \to Qr^2 \to \dots \quad .$$

The theorem may be proved more directly by applying Fitting's Lemma 311 to the morphism of congruences

$$\varphi : {}^2Q \to {}^2Q ; (x,y) \mapsto (x,x.y^r) .$$

This is left as an exercise to the reader.

3.2 Centrality and direct decompositions.

This section uses centrality to give improved versions of the Classical Exchange Theorem and Classical Unique Factorisation

Theorem, with weakened hypotheses and strengthened conclusions. To begin with, some elementary facts about centralising and direct decompositions are needed.

321 PROPOSITION (i) Let $\delta \leqslant \beta \sqcap \gamma$. If β centralises δ , then it also centralises β_γ^δ .

(ii) If $\beta \sqcap \gamma = \delta \sqcap \varepsilon$, then β centralises $\delta_\varepsilon^\gamma$.

(iii) If there is a monomorphism of congruences $\varphi : \beta \to \delta$, and γ centralises δ , then it centralises β .

Proof. (i) Let β centralise δ by means of $(\beta|\delta)$. Define a relation $(\beta|\beta_\gamma^\delta)$ on β_γ^δ by $(x,y)\ (\beta|\beta_\gamma^\delta)\ (x',y') \Leftrightarrow \exists\ z, z'$. $x\ \beta\ y\ \gamma\ z$, $x'\ \beta\ y'\ \gamma\ z'$, and $(x,z)\ (\beta|\delta)\ (x',z')$:

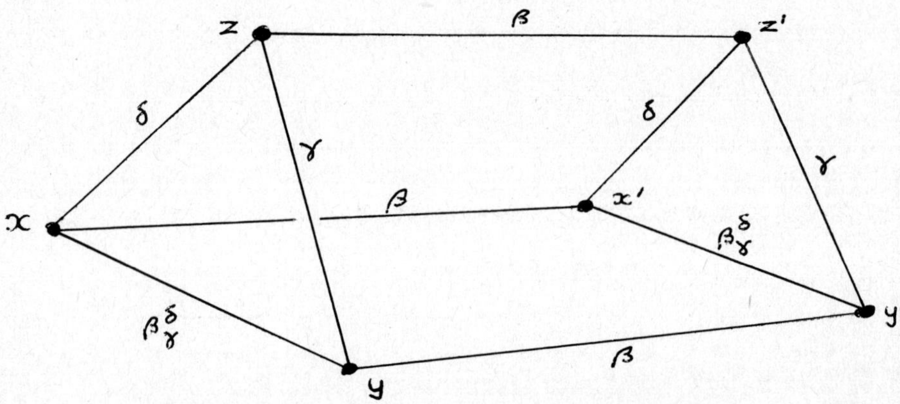

Clearly, $(\beta|\beta_\gamma^\delta)$ is a congruence on β_γ^δ satisfying (C0) of Corollary 224. If $(x,x)\ (\beta|\beta_\gamma^\delta)\ (x,y')$, then with the notation as above $x = x'$ $\Rightarrow z = z'$ by (C1) for $(\beta|\delta)$, and then $x = y'$ since $\beta \cap \gamma = \hat{A}$. Thus (C4) is also satisfied.

(ii) Define $(\beta|\delta^{\gamma}_{\varepsilon})$ on $\delta^{\gamma}_{\varepsilon}$ by $(x,y)\ (\beta|\delta^{\gamma}_{\varepsilon})\ (x',y') \Leftrightarrow x\ \beta\ x'$, $y\ \beta\ y'$, and $\exists\ z\ ,\ z'\ .\ z\ \beta\ z'\ ,\ x\ \gamma\ z\quad y\ \delta\ x$, and $x'\ \gamma\ z'\ \varepsilon\ y'\ \delta\ x'$:

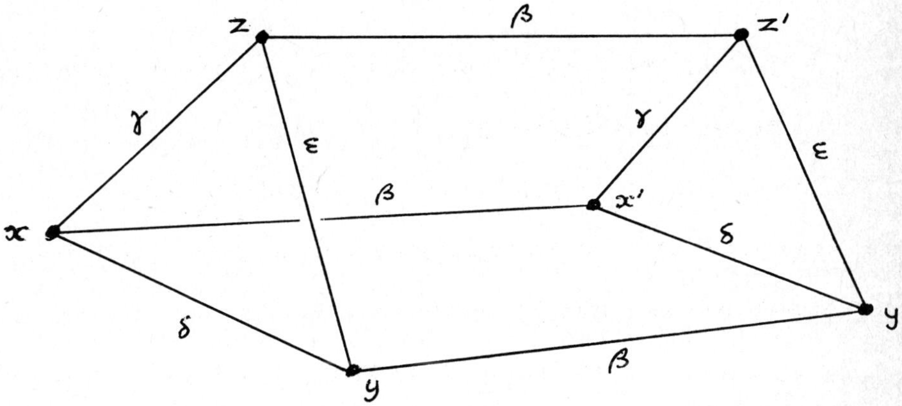

Clearly, $(\beta|\delta^{\gamma}_{\varepsilon})$ is a congruence on $\delta^{\gamma}_{\varepsilon}$ satisfying (C0) of Corollary 224. If $(x,x)\ (\beta|\delta^{\gamma}_{\varepsilon})\ (x,y')$, then with the notation as above $(z,z') \in \beta \cap \gamma = \hat{A}$, so $z = z'$, and then $(x,y') \in \delta \cap \varepsilon = \hat{A}$. Thus (C4) is also satisfied.

(iii) Define $(\gamma|\beta)$ on β by $(x,y)\ (\gamma|\beta)\ (x',y') \Leftrightarrow$ $(x,y)\varphi\ (\gamma|\delta)\ (x',y')\varphi$:

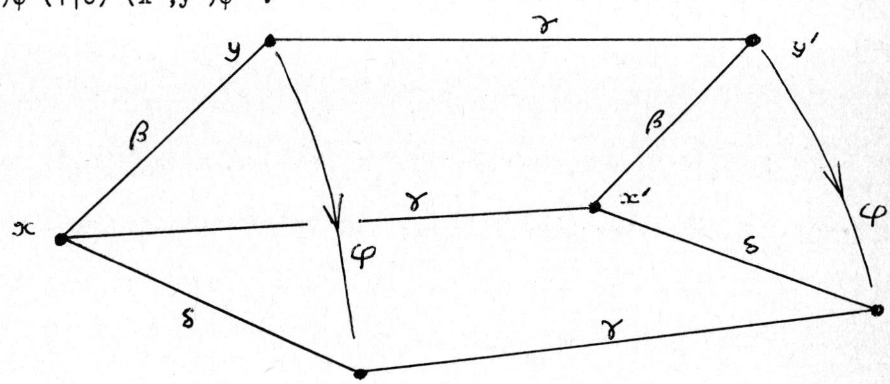

This is clearly a congruence on β , and satisfies (CO) since $\pi^0 = \varphi\pi^0$. Suppose $(x,x)\ (\gamma|\beta)\ (x,y)$. Then $(x,x)\ (\gamma|\delta)\ (x,y)\varphi$, so by (C4) for $(\gamma|\delta)$, $(x,y)\varphi = (x,x)$, i.e. $(x,y) \in \mathrm{Ker}\varphi$. But φ is a monomorphism of congruences, so $y = x$, giving (C4) for $(\gamma|\beta)$.]

322 PROPOSITION Let α , β , γ be congruences on A . If $\alpha = \beta \sqcap \gamma = \beta \sqcap \delta$, then $\beta_\gamma^\delta \leqslant \eta(\alpha)$.

Proof. Since $\beta \cap \delta = \hat{A}$, β centralises δ . Then by Proposition 321(i), β centralises β_γ^δ . But δ also centralises β_γ^δ , so by Corollary 227 $\beta \sqcap \delta = \alpha$ does.]

In Proposition 322, the isomorphism $f : \gamma \to \delta \ ; (x,y) \mapsto (x,z)$, where $x\ \gamma\ y\ \beta\ z\ \delta\ x$, is described as a central isomorphism with respect to α , since $\forall\ t \in \gamma$, $(t\pi^1, tf\pi^1) \in \alpha \cap \eta(\alpha)$. If $\alpha = {}^{2'}\!A$, f is just called a central isomorphism. Note that if $f : \beta \to \beta'$ and $g : \gamma \to \gamma'$ are central isomorphisms, then there is a central isomorphism $f \sqcap g : \beta \sqcap \gamma \to \beta' \sqcap \gamma'$ defined by $(x,z) \mapsto (x,z')$, where $x\ \beta\ y\ \gamma\ z$,

$(x,y)f = (x,y')$,

$(y,z)\ (\zeta(A)|^2A)\ (y',z'')$,

and $(y',z'')g = (y',z'')$:

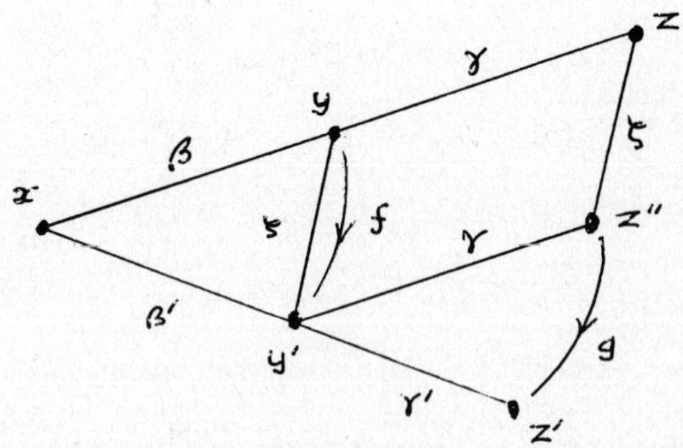

Proposition 321 and the Classical Exchange Theorem give the following, known as the Partial Exchange Theorem :

323 THEOREM If $\alpha \wedge \eta(\alpha)$ has the maximal and minimal conditions on subcongruences, and $\alpha = \beta \sqcap \gamma = \delta_1 \sqcap \cdots \sqcap \delta_n$, then for $i = 1,\ldots,n$ there are subcongruences δ_i' of δ_i such that $\alpha = \beta \sqcap \gamma = \beta \sqcap \delta_1' \sqcap \cdots \sqcap \delta_n'$.

Proof. For each i , let $\bar{\delta}_i = \delta_1 \sqcap \cdots \sqcap \delta_{i-1} \sqcap \delta_{i+1} \sqcap \cdots \sqcap \delta_n$. Then $\beta \sqcap \gamma = \delta_i \sqcap \bar{\delta}_i$. Let $\gamma_i = \delta_i \frac{\gamma}{\bar{\delta}_i}$. By Proposition 321(ii), β centralises γ_i . By Proposition 321(i), β centralises $\beta_{\gamma}^{\gamma_i}$, and since $\beta_{\gamma}^{\gamma_i} \leqslant \beta$, γ centralises $\beta_{\gamma}^{\gamma_i}$, so by Corollary 227 $\beta \sqcap \gamma = \alpha$ centralises $\beta_{\gamma}^{\gamma_i}$. Thus $\beta_{\gamma}^{\gamma_i} \leqslant \beta \wedge \eta(\alpha)$. But $\gamma_i \leqslant \beta_{\gamma}^{\gamma_i} \sqcap \gamma$, so $\gamma_i \leqslant (\beta \wedge \eta(\alpha)) \sqcap \gamma$. Since $\gamma \leqslant \gamma_1 \sqcap \cdots \sqcap \gamma_n$, it follows that

$$324 \quad \gamma \leqslant \prod_{i=1}^{n} (((\beta \wedge \eta(\alpha)) \sqcap \gamma) \wedge \delta_i) .$$

For each i , let $\delta_i^* = \delta_i \frac{\beta \wedge \eta(\alpha)}{\bar{\delta}_i}$. Then $\beta \wedge \eta(\alpha) \leqslant \delta_1^* \sqcap \cdots \sqcap \delta_n^*$. Since $\alpha = \delta_i \sqcap \bar{\delta}_i$ centralises $\beta \wedge \eta(\alpha)$, it follows from Proposition 321(i) that δ_i centralises δ_i^* . Since $\delta_i^* \leqslant \delta_i$, $\bar{\delta}_i$ centralises δ_i^* , and so by Corollary 227 $\delta_i \sqcap \bar{\delta}_i = \alpha$ centralises δ_i^* , i.e. $\delta_i^* \leqslant \eta(\alpha)$. Since $\alpha = \beta \sqcap \gamma$ centralises δ_i^* , it follows from Proposition 321(i) that β centralises $\beta_{\gamma}^{\delta_i^*}$. Since $\beta_{\gamma}^{\delta_i^*} \leqslant \beta$, γ centralises $\beta_{\gamma}^{\delta_i^*}$, and so by Corollary 227 $\beta \sqcap \gamma = \alpha$ centralises $\beta_{\gamma}^{\delta_i^*}$, i.e. $\beta_{\gamma}^{\delta_i^*} \leqslant \beta \wedge \eta(\alpha)$. Thus $\delta_i^* \leqslant$

$\beta_{\gamma}^{\delta_i^*} \sqcap \gamma \leqslant ((\beta \frown \eta(\alpha)) \sqcap \gamma) \frown \delta_i$. Hence

$$\beta \frown \eta(\alpha) \leqslant \prod_{i=1}^{n} (((\beta \frown \eta(\alpha)) \sqcap \gamma) \frown \delta_i) .$$

From 324,

325 $$(\beta \frown \eta(\alpha)) \sqcap \gamma = \prod_{i=1}^{n} (((\beta \frown \eta(\alpha)) \sqcap \gamma) \frown \delta_i) .$$

Since $\alpha \frown \eta(\alpha)$ has the minimal condition on subcongruences, $\beta \frown \eta(\alpha) = \beta_1 \sqcap \cdots \sqcap \beta_p$ for some finite number p of indecomposable β_j . It will be shown by induction on p that

326 $$\exists\, \delta_i' \leqslant \delta_i \,.\, (\beta \frown \eta(\alpha)) \sqcap \gamma = (\beta \frown \eta(\alpha)) \sqcap \delta_1' \sqcap \cdots \sqcap \delta_n' .$$

This is sufficient to give the result, for, momentarily letting $\varepsilon = \delta_1' \sqcap \cdots \sqcap \delta_n'$, $(\beta \frown \eta(\alpha)) \sqcap \gamma = (\beta \frown \eta(\alpha)) \sqcap \varepsilon$. Suppose $(x,y) \in \beta \sqcap \gamma$. Then $\exists\, t \,.\, x \,\beta\, t \,\gamma\, y$. Now $(t,y) \in (\beta \frown \eta(\alpha)) \sqcap \varepsilon$. $\Rightarrow \exists\, u \,.\, t \,(\beta \frown \eta(\alpha))\, u \,\varepsilon\, y$. $\Rightarrow (x,y) \in \beta \circ \varepsilon$. Hence $\beta \circ \varepsilon = \alpha$. Suppose $(x,y) \in \beta \frown \varepsilon$. In particular $(x,y) \in \varepsilon \leqslant (\beta \frown \eta(\alpha)) \sqcap \gamma$. So $\exists\, t \,.\, x \,(\beta \frown \eta(\alpha))\, t \,\gamma\, y$. Then $(t,y) \in \beta \frown \gamma = \hat{A}$. $\Rightarrow t = y$. $\Rightarrow (x,y) \in (\beta \frown \eta(\alpha)) \frown \varepsilon = \hat{A}$. Hence $\beta \frown \varepsilon = \hat{A}$. Thus $\alpha = \beta \sqcap \varepsilon$.

It thus remains to prove 326 by induction on p . If $p = 0$, $\beta \frown \eta(\alpha) = \hat{A}$. Equation 326 is then just 325 with $\delta_i' = \gamma \frown \delta_i$. Suppose 326 holds for $\beta \frown \eta(\alpha)$ having an expression as a product of less than p indecomposables. Now $(\beta \frown \eta(\alpha)) \sqcap \gamma = \delta_{n+1}'' \sqcap (\beta_p \sqcap \gamma)$, where $\delta_{n+1}'' = \beta_1 \sqcap \cdots \sqcap \beta_{p-1}$. By induction $\exists\, \delta_i'' \leqslant \delta_i$.

$(\beta \cap \eta(\alpha)) \sqcap \gamma = \delta_1'' \sqcap \cdots \sqcap \delta_n'' \sqcap \delta_{n+1}''$, i.e. $\beta_p \sqcap (\delta_{n+1}'' \sqcap \gamma) =$ $\delta_1'' \sqcap \cdots \sqcap \delta_{n+1}''$. Note that β_p is indecomposable, and since it is contained in $\alpha \cap \eta(\alpha)$ it has the maximal and minimal conditions on subcongruences. The Classical Exchange Theorem 315 then shows $\exists\ 1 \leqslant l \leqslant n$. $\delta_l'' = \xi \sqcap \varepsilon$ and $(\beta \cap \eta(\alpha)) \sqcap \gamma = \xi \sqcap \delta_{n+1}'' \sqcap \gamma =$ $\beta_p \sqcap \varepsilon \sqcap \delta_1'' \sqcap \cdots \sqcap \delta_{l-1}'' \sqcap \delta_{l+1}'' \sqcap \cdots \sqcap \delta_{n+1}''$. Note $l \neq n+1$ since $\xi \leqslant \delta_l''$ and $\xi \cap \delta_{n+1}'' = \hat{A}$. Set $\delta_l' = \varepsilon$, and $\delta_i' = \delta_i''$ if $l \neq i \leqslant n$. Then $(\beta \cap \eta(\alpha)) \sqcap \gamma = (\beta \cap \eta(\alpha)) \sqcap \delta_1' \sqcap \cdots \sqcap \delta_n'$. 326 thus holds for all p .]

Now for the full <u>Exchange Theorem</u> :

327 THEOREM If β is indecomposable, $\alpha \cap \eta(\alpha)$ has the maximal and minimal conditions on subcongruences, and $\alpha = \beta \sqcap \gamma =$ $\delta_1 \sqcap \cdots \sqcap \delta_n$, then $\exists\ 1 \leqslant l \leqslant n$. $\delta_l = \xi \sqcap \varepsilon$ and $\beta \sqcap \gamma = \xi \sqcap \gamma =$ $\beta \sqcap \varepsilon \sqcap \delta_1 \sqcap \cdots \sqcap \delta_{l-1} \sqcap \delta_{l+1} \sqcap \cdots \sqcap \delta_n$.

<u>Proof.</u> If $\beta \leqslant \eta(\alpha)$, the result follows by the Classical Exchange Theorem 315. So assume $\beta \not\leqslant \eta(\alpha)$. By the Partial Exchange Theorem 323, $\exists\ \delta_i' \leqslant \delta_i$. $\beta \sqcap \gamma = \beta \sqcap \delta_1' \sqcap \cdots \sqcap \delta_n'$. By the modular law in the lattice of congruences, $\delta_i = (\beta \sqcap \delta_1' \sqcap \cdots \sqcap \delta_n') \cap \delta_i =$ $\delta_i' \sqcap ((\beta \sqcap \delta_1' \sqcap \cdots \sqcap \delta_{i-1}' \sqcap \delta_{i+1}' \sqcap \cdots \sqcap \delta_n') \cap \delta_i)$. Let $\delta_i'' =$ $(\beta \sqcap \delta_1' \sqcap \cdots \sqcap \delta_{i-1}' \sqcap \delta_{i+1}' \sqcap \cdots \sqcap \delta_n') \cap \delta_i$. Then

$\beta \sqcap \delta_1' \sqcap \cdots \sqcap \delta_n' = \delta_1'' \sqcap \cdots \sqcap \delta_n'' \sqcap \delta_1' \sqcap \cdots \sqcap \delta_n'$, so β is isomorphic with $\delta_1'' \sqcap \cdots \sqcap \delta_n''$. But β is indecomposable, and so there is just one i , say $i = l$, with $\delta_i'' > \hat{A}$. Hence $\delta_i' = \delta_i$ for $l \neq i = 1,\ldots,n$. Putting $\xi = \delta_l''$ and $\varepsilon = \delta_l'$, $\alpha = \beta \sqcap \gamma =$ $\beta \sqcap \varepsilon \sqcap \delta_1 \sqcap \cdots \sqcap \delta_{l-1} \sqcap \delta_{l+1} \sqcap \cdots \sqcap \delta_n =$ $\xi \sqcap \varepsilon \sqcap \delta_1 \sqcap \cdots \sqcap \delta_{l-1} \sqcap \delta_{l+1} \sqcap \cdots \sqcap \delta_n$. In particular, β is isomorphic with ξ . The Partial Exchange Theorem 323 then yields $\beta' \leqslant \beta$, $\gamma' \leqslant \gamma$ such that $\beta \sqcap \gamma = \xi \sqcap \beta' \sqcap \gamma'$. Let $\beta'' =$ $(\xi \sqcap \gamma') \cap \beta$, $\gamma'' = (\xi \sqcap \beta') \cap \gamma$. Then $\beta = (\beta' \sqcap (\xi \sqcap \gamma')) \cap \beta =$ $\beta' \sqcap ((\xi \sqcap \gamma') \cap \beta) = \beta' \sqcap \beta''$ and $\gamma = (\gamma' \sqcap (\xi \sqcap \beta')) \cap \gamma =$ $\gamma' \sqcap ((\xi \sqcap \beta') \cap \gamma) = \gamma' \sqcap \gamma''$ by the modular law in the lattice of congruences, so $\beta \sqcap \gamma = \xi \sqcap \beta' \sqcap \gamma' = \beta'' \sqcap \gamma'' \sqcap \beta' \sqcap \gamma'$. Thus ξ becomes isomorphic with $\beta'' \sqcap \gamma''$, and so β is isomorphic with $\beta'' \sqcap \gamma''$. But β is indecomposable, and so $\beta'' = \hat{A}$ or $\gamma'' = \hat{A}$. If $\beta'' = \hat{A}$, there is an isomorphism $f : \beta \to \gamma''$. Now $\beta \cap \gamma = \hat{A}$, and $\gamma'' \leqslant \gamma$, so β centralises γ'' and γ . Then by Proposition 321 (iii) β centralises β , whence by Corollary 227 $\alpha = \beta \sqcap \gamma$ centralises β , i.e. $\beta \leqslant \eta(\alpha)$. This is a contradiction to the initial assumption. The only alternative is that $\gamma'' = \hat{A}$ and $\beta'' > \hat{A}$. But the indecomposable $\beta = \beta' \sqcap \beta''$. Hence $\beta' = \hat{A}$. Also $\gamma = \gamma'$. Thus $\alpha = \beta \sqcap \gamma = \xi \sqcap \gamma$, as required.]

Finally, the Unique Factorisation Theorem :

328 THEOREM If $\alpha \cap \eta(\alpha)$ has the maximal and minimal conditions on subcongruences, then α has at most one decomposition $\alpha = \alpha_1 \sqcap \cdots \sqcap \alpha_m$ into finitely many indecomposable factors. More precisely, if $\alpha = \alpha_1 \sqcap \cdots \sqcap \alpha_m = \beta_1 \sqcap \cdots \sqcap \beta_n$ with the α_i and β_j indecomposable, then $m = n$ and there is a permutation σ of $\{1,\ldots,m\}$ such that $\alpha = \beta_{\sigma(1)} \sqcap \cdots \sqcap \beta_{\sigma(r)} \sqcap \alpha_{r+1} \sqcap \cdots \sqcap \alpha_m$ for $r = 0,\ldots,m$. Further, α_i is centrally isomorphic with $\beta_{\sigma(i)}$ with respect to α for $i = 1,\ldots,m$.

Proof. This goes by induction on r just as for the Classical Unique Factorisation Theorem 316, but using the Exchange Theorem 327 instead of the Classical Exchange Theorem 315. The last statement follows from Proposition 322.]

Schematically, the central isomorphisms (with respect to α) of the Unique Factorisation Theorem may be represented as follows :

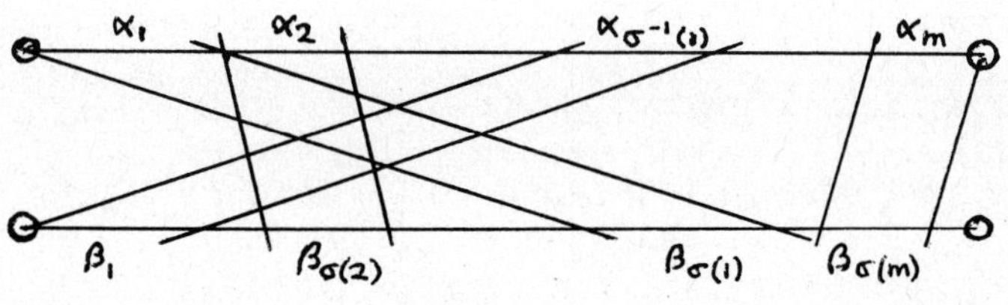

This sort of diagram will be used in the proof of the Cancellation Theorem 424.

NOTES

The methods of this chapter are analogous to those of B. Jónsson and A. Tarski [JT]. Define a Jónsson-Tarski variety to be one having a binary operation + and a nullary operation 0 such that $0+x = x$ and $x+0 = x$ are identities. For a Jónsson-Tarski variety, results like those of Chapters 2 and 3 for Mal'cev varieties hold, but with (subtractive [JT, Defn. 1.15]) subalgebras replacing congruences. The first aim of a categorical approach to the material of these notes should be a unified treatment of Jónsson-Tarski and Mal'cev varieties.

Before the publication of Mal'cev's paper [Ma], Goldie [Go] studied direct decompositions of pointed Mal'cev algebras. However, defining $x+y = (x,0,y)P$ reduces such algebras to the Jónsson-Tarski case. This was (chronologically) the first example of the power of Mal'cev's Theorem 142.

4 Central isotopy and cancellation

The previous chapter was concerned with direct decompositions of a single Mal'cev algebra and congruences on it. The subject of this chapter is the behaviour of a whole Mal'cev variety $\underline{\underline{T}}$ under the direct product. The first section introduces the concept of central isotopy. This is an extremely important equivalence relation on $\underline{\underline{T}}$. For many purposes central isotopy classes of $\underline{\underline{T}}$-algebras are more fundamental than isomorphism classes. The second section gives the major example of this : it is shown that cancellation under the direct product holds only up to central isotopy, rather than up to isomorphism. The third section gives another example.

4.1 Central isotopy.

Two $\underline{\underline{T}}$-algebras (A,ρ) , (B,σ) are said to be

<u>isotopic</u>, written $(A,\rho) \sim (B,\sigma)$, if there are, for each ω in Ω (say of arity n), bijections $\theta_0, \theta_1, \ldots, \theta_n : A \to B$ such that $x_1\theta_1 \ldots x_n\theta_n\sigma(\omega) = x_1 \ldots x_n\rho(\omega)\theta_0$. (B,σ) is said to be an <u>isotope</u> of (A,ρ). Isotopy is a generalisation of isomorphism - (A,ρ) is isomorphic to (B,σ) iff all the θ_i for all the ω are the same. Note that isotopy is an equivalence relation on $\underline{\underline{T}}$-algebras. (B,σ) is said to be a <u>central isotope</u> of (A,ρ) iff there is a bijection $\theta : B \to A$, called a <u>central shift</u>, such that for each ω in Ω (say of arity n),

411 $\exists\, (a_\omega,\bar{a}_\omega) \in \zeta(A)\ .\ \forall\, b_1, \ldots, b_n \in B$,

$$(a_\omega,\bar{a}_\omega)\ ({}^2A|\zeta(A))\ (b_1 \ldots b_n\sigma(\omega)\theta, b_1\theta \ldots b_n\theta\rho(\omega))\ .$$

If this holds, write $(B,\sigma) \simeq (A,\rho)$. Note that by (C1) for $\zeta(A)$, $(B,\sigma) \simeq (A,\rho) \Rightarrow (B,\sigma) \sim (A,\rho)$.

412 PROPOSITION $\simeq$ is an equivalence relation. Further, if $\theta : B \to A$ is a central shift, then $\zeta(A) = \zeta(B)^2\theta$ and $({}^2A|\zeta(A)) = ({}^2B|\zeta(B))^4\theta$.

<u>Proof.</u> Throughout, let ω be an n-ary operation. If $b_i\theta\ \zeta(A)\ b_i'\theta$ for $i = 1,\ldots,n$, then $b_i\sigma(\omega)\theta\ \zeta(A)\ b_i\theta\rho(\omega)\ \zeta(A)\ b_i'\theta\rho(\omega)\ \zeta(A)\ b_i'\sigma(\omega)\theta$, so $\zeta(A)^2\theta^{-1} \leq {}^2(B,\sigma)$. Since θ bijects, $\hat{B} \subseteq \zeta(A)^2\theta^{-1}$, and so $\zeta(A)^2\theta^{-1}$ is a congruence on (B,σ). If $(b_i\theta, b_i'\theta)\ ({}^2A|\zeta(A))\ (c_i\theta, c_i'\theta)$ for $i = 1,\ldots,n$, then

$(b_i\sigma(\omega)\theta, b_i\theta\rho(\omega))\ ({}^2A|\zeta(A))\ (a_\omega,\bar{a}_\omega)\ ({}^2A|\zeta(A))\ (c_i\sigma(\omega)\theta, c_i\theta\rho(\omega))$,

$(b_i\theta\rho(\omega), b_i'\theta\rho(\omega))\ ({}^2A|\zeta(A))\ (c_i\theta\rho(\omega), c_i'\theta\rho(\omega))$, and

$(b_i'\theta\rho(\omega), b_i'\sigma(\omega)\theta)\ ({}^2A|\zeta(A))\ (\bar{a}_\omega, a_\omega)\ ({}^2A|\zeta(A))\ (c_i'\theta\rho(\omega), c_i'\sigma(\omega)\theta)$.

Since $({}^2A|\zeta(A))$ respects transitivity of $\zeta(A)$, $(b_i\sigma(\omega)\theta, b_i'\sigma(\omega)\theta)$ $({}^2A|\zeta(A))\ (c_i\sigma(\omega)\theta, c_i'\sigma(\omega)\theta)$. Thus $({}^2A|\zeta(A))^4\theta^{-1} \leqslant {}^2(\zeta(A)^2\theta^{-1})$. Since θ bijects, the set-theoretic properties $\widehat{\zeta(A)^2\theta^{-1}} \subseteq$ $({}^2A|\zeta(A))^4\theta^{-1}$ and (C0) , (C4) of Corollary 224 hold, so ${}^2B$ centralises $\zeta(A)^2\theta^{-1}$ by means of $({}^2A|\zeta(A))^4\theta^{-1}$. By 228 and the definition of $\zeta(B)$, $\zeta(A)^2\theta^{-1} \leqslant \zeta(B)$. By Proposition 221 $({}^2A|\zeta(A))^4\theta^{-1} = ({}^2B|\zeta(B)) \cap {}^2(\zeta(A)^2\theta^{-1})$.

With the notation of 411, let $b_\omega = \bar{a}_\omega\theta^{-1}$, $\bar{b}_\omega = a_\omega\theta^{-1}$. Now $\forall\ a_1, \ldots, a_n \in A$, $(\bar{b}_\omega\theta, b_\omega\theta)\ ({}^2A|\zeta(A))\ (a_i\theta^{-1}\sigma(\omega)\theta, a_i\rho(\omega))$. Applying ${}^2\theta^{-1}$ and using the respect of $({}^2A|\zeta(A))$ for the symmetry of $\zeta(A)$, $(b_\omega,\bar{b}_\omega)\ ({}^2A|\zeta(A))^4\theta^{-1}\ (a_i\rho(\omega)\theta^{-1}, a_i\theta^{-1}\sigma(\omega))$. From above, this can be written as $(b_\omega,\bar{b}_\omega)\ ({}^2B|\zeta(B))\ (a_i\rho(\omega)\theta^{-1}, a_i\theta^{-1}\sigma(\omega))$. Thus $\theta^{-1} : A \rightarrow B$ is also a central shift. In particular, the relation $\simeq$ is symmetric. Repeating the above procedure for the new central shift θ^{-1} , $\zeta(B)^2\theta \leqslant \zeta(A)$. Hence $\zeta(A) = \zeta(B)^2\theta$ and $({}^2A|\zeta(A)) = ({}^2B|\zeta(B))^4\theta$.

Let $\varphi : C \rightarrow B$ also be a central shift making (C,τ) $\simeq (B,\sigma)$, say with $(b_\omega', b_\omega^*)\ ({}^2B|\zeta(B))\ (c_i\tau(\omega)\varphi, c_i\varphi\sigma(\omega))$. Applying θ

and using the above, $(b'_\omega e, b^*_\omega)\ ({}^2A|\zeta(A))\ (c_i\tau(\omega)\varphi e, c_i\varphi\sigma(\omega)e)$. By property (C1) for $({}^2A|\zeta(A))$, there is a unique a'_ω such that $(a'_\omega, a_\omega)({}^2A|\zeta(A))(b'_\omega e, b^*_\omega)$. Then $(a'_\omega, a_\omega)({}^2A|\zeta(A))(c_i\tau(\omega)\varphi e, c_i\varphi\sigma(\omega)e)$ and $(a_\omega, \bar{a}_\omega)\ ({}^2A|\zeta(A))\ (c_i\varphi\sigma(\omega)e, c_i\varphi e\rho(\omega))$ imply $(a'_\omega, \bar{a}_\omega)\ ({}^2A|\zeta(A))$ $(c_i\tau(\omega)\varphi e, c_i\varphi e\rho(\omega))$ by the respect of $({}^2A|\zeta(A))$ for the transitivity of $\zeta(A)$. Thus $\simeq$ is a transitive relation. Since 1_A is a central shift (with $a_\omega = \bar{a}_\omega$ for all ω), $\simeq$ is reflexive. Thus $\simeq$ is an equivalence relation.]

Immediate corollaries of this are that central isotopes of $\underline{\underline{Z}}$-algebras are $\underline{\underline{Z}}$-algebras, and that $B \times C \simeq B' \times C'$ if $B \simeq B'$ and $C \simeq C'$.

413 Any isomorphism $e : (B,\sigma) \to (A,\rho)$ is a central shift, with $a_\omega = \bar{a}_\omega$ for all ω . As a partial converse, if (A,ρ) , (B,σ) have respective singleton subalgebras $(\{e\},\rho)$, $(\{f\},\sigma)$, and if $e : B \to A$ is a central shift with $fe = e$, then e is an isomorphism of (A,ρ) and (B,σ) .

414 If (A,ρ) has the singleton subalgebra $(\{e\},\rho)$, then the mapping $\zeta(A)^{({}^2A|\zeta(A))} \to e^{\zeta(A)}$; $(x,y)^{({}^2A|\zeta(A))} \mapsto$

$$(\{e\}(\pi^0)^{-1} \cap (x,y)^{({}^2A|\zeta(A))})\pi^1$$

is an isomorphism of $\underline{\underline{T}}$-algebras. By Proposition 222, $\zeta(A)^{({}^2A|\zeta(A))}$ is an abelian group in $\underline{\underline{T}}$, and so by this isomorphism $e^{\zeta(A)}$ becomes one too, with identity e . Singleton subalgebras within $e^{\zeta(A)}$ form a subgroup G . Central shifts of (A,ρ) to itself permuting singleton subalgebras within a given $\zeta(A)$-class then form an abelian group of fixed-point-free automorphisms of (A,ρ) isomorphic with G .

There is a fundamental connection between central isomorphism of congruences on an algebra and central isotopy of quotient algebras as follows:

415 PROPOSITION Suppose ${}^2A = \beta \sqcap \gamma = \delta \sqcap \varepsilon$, and there is a central isomorphism of congruences $\varphi : \beta \to \delta$. Then A^{ε} is a central isotope of A^{γ} .

<u>Proof.</u> Fix $a \in A$. Define $\theta : A^{\gamma} \to A$; $b^{\gamma} \mapsto d^{\varepsilon}$, where $\varphi : (a,b) \mapsto (a,d)$.

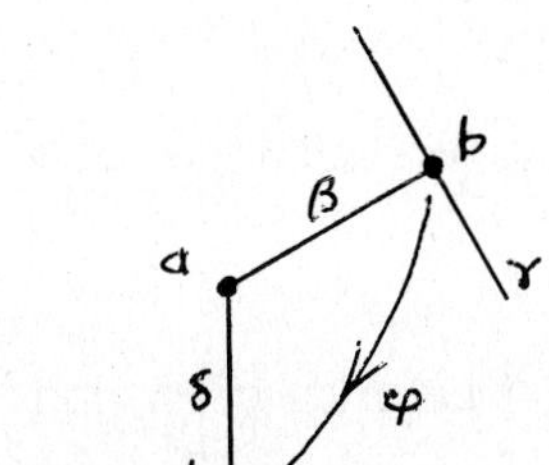

θ clearly bijects. It will be shown that it is a central shift.

γ centralises β , so by Proposition 321(iii) it centralises δ . By symmetry δ centralises γ , and then by Proposition 321(i) it centralises $\delta_{\varepsilon}^{\gamma}$. ε also centralises $\delta_{\varepsilon}^{\gamma}$, so by Corollary 227 ${}^{2}A = \delta \sqcap \varepsilon$ does, i.e. $\delta_{\varepsilon}^{\gamma} \leqslant \zeta(A)$.

Let ω be an n-ary operation. Suppose that for $i = 1,\ldots,n$, $(a,b_i)\varphi = (a,d_i)$, i.e. $b_i^{\gamma}\theta = d_i^{\varepsilon}$. Define a' by

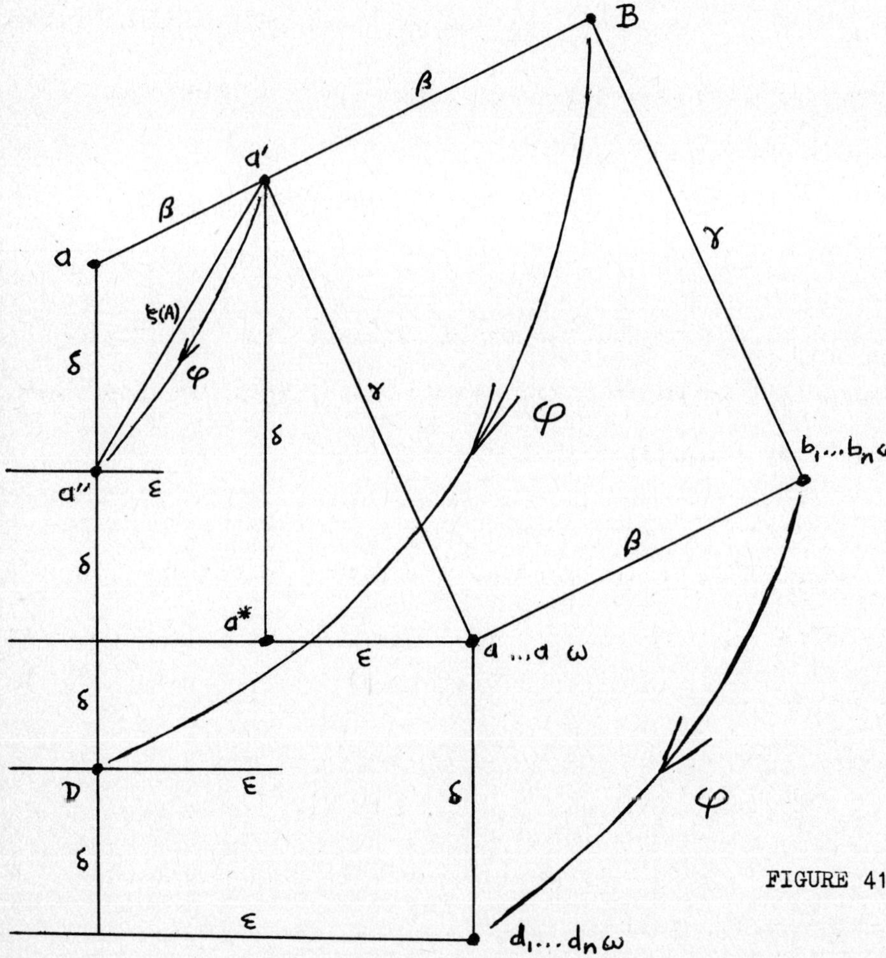

FIGURE 416

$a\ \beta\ a'\ \gamma\ a\ldots a\omega$, a^* by $a'\ \delta\ a^*\ \varepsilon\ a\ldots a\omega$, a'' by $(a,a')\varphi = (a,a'')$, B by $(a',B)\ (\gamma|\beta)\ (a\ldots a\omega, b_1\ldots b_n\omega)$, and D by $(a,B)\varphi = (a,D)$ (Figure 416). $(a^*,a') \in \delta_\varepsilon^\gamma \leqslant \zeta(A)$, and since φ is a central isomorphism, $(a'',a') \in \zeta(A)$. Thus $(a'',a^*) \in \zeta(A)$, so by Proposition 231(ii) $(a''^\varepsilon, a\ldots a\omega^\varepsilon) = (a''^\varepsilon, a^{*\varepsilon}) \in \zeta(A^\varepsilon)$. Now

$$(a,B) = ((a,a'),(a\ldots a\omega,a\ldots a\omega),(a\ldots a\omega,b_1\ldots b_n\omega))P$$

$\Rightarrow$ $(a,D) = ((a,a''),(a\ldots a\omega,a\ldots a\omega),(a\ldots a\omega,d_1\ldots d_n\omega))P$. Thus $D = (a'',a\ldots a\omega,d_1\ldots d_n\omega)P$, whence $(a''^\varepsilon, a\ldots a\omega^\varepsilon)\ ({}^2A^\varepsilon|\zeta(A^\varepsilon))$ $(D^\varepsilon, d_1\ldots d_n\omega^\varepsilon) = (b_1\ldots b_n\omega^\gamma\theta, b_1^\gamma\theta\ldots b_n^\gamma\theta\omega)$. This shows that θ is indeed a central shift.]

The full generality of Proposition 415 will not be needed until the proof of the Cancellation Theorem 424. The next result uses a weak form of Proposition 415.

417 PROPOSITION Let A be a $\underline{\underline{Z}}$-algebra. Then there is a $\underline{\underline{Z}}_0$-algebra B such that $B \times A \cong A \times A$. Further, B is a central isotope of A .

Proof. In the notation of Proposition 229(i), with $\gamma = \beta = {}^2A$, ${}^4A = E^0 \sqcap ({}^2A|{}^2A) = E^0 \sqcap E^1$. Note that $({}^2A)\mathrm{nat}E^i \cong A$ for $i = 0,1$. By 233, $({}^2A)\mathrm{nat}({}^2A|{}^2A)$ is in $\underline{\underline{Z}}$, with singleton subalgebra $\{\hat{A}\}$. Let $B = ({}^2A)\mathrm{nat}({}^2A|{}^2A)$. Then $A \times B \cong A \times A$. Further, since the identity mapping from E^0 to itself is a central isomorphism, Proposition 415 shows that B is a central isotope of A .]

It is instructive to work through Proposition 417 taking $A = \{1,2,3\}$ to be the quasigroup with multiplication table

.	1	2	3
1	2	1	3
2	1	3	2
3	3	2	1

.

Collecting Proposition 417 and other miscellaneous results gives the following Structure Theorem for $\underline{\underline{Z}}$-algebras :

418 THEOREM A $\underline{\underline{T}}$-algebra A is in $\underline{\underline{Z}}$ iff it is a central isotope of a $\underline{\underline{Z}}_0$-algebra. $\underline{\underline{Z}}_0$-algebras are just abelian groups in the category $\underline{\underline{T}}$.

Proof. By Proposition 412, any central isotope of a $\underline{\underline{Z}}$-algebra is a $\underline{\underline{Z}}$-algebra. By Proposition 417, every $\underline{\underline{Z}}$-algebra is a central isotope of a $\underline{\underline{Z}}_0$-algebra. By 414, every $\underline{\underline{Z}}_0$-algebra is an abelian group in the category $\underline{\underline{T}}$. By Proposition 234, the variety of abelian groups in $\underline{\underline{T}}$ is contained in $\underline{\underline{Z}}$.]

419 COROLLARY If $\theta : B \to C$ is a central shift in $\underline{\underline{T}}$ from one $\underline{\underline{Z}}_0$-algebra to another, then it is an isomorphism.

Proof. Let $\{e\}$, $\{f\}$ respectively be singleton subalgebras of B and C . Let $(C,+)$ be the abelian group with identity f given by 414. Suppose $e\theta = g$. Since θ is a central shift, there is for each operation ω (of arity n , say) an element a of C such that $b_1 \ldots b_n \omega\theta = b_1\theta \ldots b_n\theta\omega + a_\omega$. Then $(b_1\theta - g)\ldots(b_n\theta - g)\omega = b_1\theta\ldots b_n\theta\omega - g\ldots g\omega = b_1\ldots b_n\omega\theta - (a_\omega + g\ldots g\omega)$, so that $b \mapsto b\theta - g$ is also a central shift. But $e\theta - g = f$, so by 413 $b \mapsto b\theta - g$ is an isomorphism.]

4.2 Cancellation.

For a Mal'cev variety $\underline{\underline{T}}$, let $[\underline{\underline{T}}]$ denote the set of isomorphism classes of finite $\underline{\underline{T}}$-algebras. $[\underline{\underline{T}}]$ is a commutative monoid under the direct product x with the class [1] of the one-element $\underline{\underline{T}}$-algebra as identity. Let $K[\underline{\underline{T}}]$ be the Grothendieck group of the monoid $[\underline{\underline{T}}]$, i.e. a group with monoid-morphism $[\underline{\underline{T}}] \to K[\underline{\underline{T}}]$ universal among monoid-morphisms from $[\underline{\underline{T}}]$ to groups. Let $[A]$ denote the image of the isomorphism class of the $\underline{\underline{T}}$-algebra A under this morphism. For two finite $\underline{\underline{T}}$-algebras C , D , $[C] = [D]$ iff there is a third $\underline{\underline{T}}$-algebra B such that $B \times C \cong B \times D$. As Proposition 417 shows, $[C] = [D]$ does not imply in general that $C \cong D$. This section gives precisely how C and D are related if $[C] = [D]$.

The first result shows that if $B \simeq C$, then $[B] = [C]$. It is rather more general, not being restricted to finite algebras.

421 THEOREM If (B,σ) is a central isotope of (C,τ) , then there is a $\underline{\underline{Z}}$-algebra Z such that $Z \times B \cong Z \times C$. Further, if B is finite (so that C is) then Z is also.

Proof. Throughout, let ω be an n-ary operation. Let $V = ({}^2B|\zeta(B))$. Suppose $e : C \to B$ is a central shift, so that $(b_\omega,\overline{b}_\omega)\ V\ (c_i\tau(\omega)e, c_ie\sigma(\omega))$. Define an action τ of Ω on B by $(b_\omega,\overline{b}_\omega)\ V\ (b_i\tau(\omega), b_i\sigma(\omega))$, using property (C1) of V . Then $e : (C,\tau) \to (B,\tau)$ is an isomorphism. Thus $(C,\tau) \cong (B,\tau) \simeq (B,\sigma)$, and the identity mapping on B is a central shift from (B,τ) to (B,σ) .

Now $\zeta(B)$ is a subset of ${}^2B$. Let $(b_i,b_i') \in \zeta(B)$. Since $(\zeta(B),{}^2\sigma)$ is a subalgebra of $({}^2B,{}^2\sigma)$, $(b_i\sigma(\omega),b_i'\sigma(\omega)) \in \zeta(B)$. By definition of the action τ on B , $(b_i'\sigma(\omega),b_i'\tau(\omega)) \in$

$\zeta(B)$. Hence by the transitivity of $\zeta(B)$, $(b_i\sigma(\omega), b_i'\tau(\omega)) \in \zeta(B)$. Thus $(\zeta(B),(\sigma,\tau))$ is a subalgebra of $({}^2B,(\sigma,\tau))$. Let $(A,\rho) = (\zeta(B),(\sigma,\tau))$. Let $K^i = \ker(\ \pi^i : \zeta(B) \to B \ ; \ (x_0,x_1) \mapsto x_i\ )$ for $i = 0,1$. The K^i are congruences on (A,ρ) , with $(A\text{nat}K^0,\rho) \cong (B,\sigma)$ and $(A\text{nat}K^1,\rho) \cong (B,\tau) \cong (C,\tau)$. Suppose $(a_i,a_i')\ V\ (b_i,b_i')$. Then $(a_i\sigma(\omega),a_i'\sigma(\omega))\ V\ (b_i\sigma(\omega),b_i'\sigma(\omega))$ and $(a_i'\sigma(\omega),a_i'\tau(\omega))\ V\ (\overline{b}_\omega,b_\omega)\ V\ (b_i'\sigma(\omega),b_i'\tau(\omega))$, so by the respect of V for the transitivity of $\zeta(B)$, $(a_i\sigma(\omega),a_i'\tau(\omega))\ V\ (b_i\sigma(\omega),b_i'\tau(\omega))$. Clearly $\hat{A} = \widehat{\zeta(B)} \subseteq V$. Thus V is a congruence on A .

In Proposition 229(i), put $A = (B,\sigma)$, $\beta = \zeta(B)$, $\gamma = {}^2B$, $(\gamma|\beta) = V$, and use the definitions and notations of the proposition and its proof. In particular, put $E^0 = E^1 = {}^2(A^V)$. The proposition says that $(A^V,{}^2\sigma)$ is a $\underline{\underline{Z}}$-algebra. Now for $(x_i,y_i)^V$ in A^V , $(y_i\tau(\omega),y_i\sigma(\omega))\ V\ (b_\omega,\overline{b}_\omega)$ implies that $((x_i\sigma(\omega),y_i\tau(\omega))^V,(x_i\sigma(\omega),y_i\sigma(\omega))^V)\ (E^1|E^0)\ ((b_\omega,b_\omega)^V,(b_\omega,\overline{b}_\omega)^V)$. Thus (A^V,ρ) is a central isotope of the $\underline{\underline{Z}}$-algebra $(A^V,{}^2\sigma)$, whence by Proposition 412 (A^V,ρ) is itself a $\underline{\underline{Z}}$-algebra. Let $(A^V,\rho) = Z$. Note that B finite $\Rightarrow$ (B) finite $\Rightarrow$ Z finite.

By the properties (C1) and (RS) of V , ${}^2A = K^0 \sqcap V = K^1 \sqcap V$. Thus $A \cong A\text{nat}V \times A\text{nat}K^0 \cong A\text{nat}V \times A\text{nat}K^1$, i.e. $A \cong Z \times B \cong Z \times C$, as required.]

To see what $[B] = [C]$ implies, it is necessary to study cancellation in $\underline{\underline{T}}_0$ first. The following, known as the <u>Pointed Exchange Theorem</u>, helps with this.

422 THEOREM Let A be a finite $\underline{\underline{T}}_0$-algebra, pointed by $\{0\} \leqslant A$. Let $B, C, D_i \leqslant A$, for $i = 1,\ldots,n$, with B indecomposable, and $B \times C = D_1 \times D_2 \times \ldots \times D_n$. Then $\exists\ 1 \leqslant l \leqslant n$. $D_l = X \times Y$ and $B \times C = X \times Y = B \times Y \times D_1 \times \ldots \times D_{l-1} \times D_{l+1} \times \ldots \times D_n$.

<u>Proof.</u> If $B \times C = S_0 \times S_1$, let $\pi^i : S_0 \times S_1 \to S_i$ denote the projections for $i = 0,1$. Then $S_i = 0^{\ker\pi^i}$. Thus there are congruences β, γ, δ_i on $B \times C$ for $i = 1,\ldots,n$ such that $\beta \sqcap \gamma = \delta_1 \sqcap \delta_2 \sqcap \ldots \sqcap \delta_n$ with $B = 0^\beta$, $C = 0^\gamma$, $D_i = 0^{\gamma_i}$, and β indecomposable ($\beta = \beta' \sqcap \beta'' \Rightarrow B \cong 0^{\beta'} \times 0^{\beta''}$). By the Classical Exchange Theorem 315, $\exists\ 1 \leqslant l \leqslant n$. $\delta_l = \xi \sqcap \varepsilon$ and $\beta \sqcap \gamma = \xi \sqcap \gamma = \beta \sqcap \varepsilon \sqcap \delta_1 \sqcap \ldots \sqcap \delta_{l-1} \sqcap \delta_{l+1} \sqcap \ldots \sqcap \delta_n$. Putting $X = 0^\xi$ and $Y = 0^\varepsilon$ gives the result.]

The Pointed Exchange Theorem is used for the following, the <u>Pointed Cancellation Theorem</u> :

423 THEOREM Let A be a finite $\underline{\underline{T}}_0$-algebra, with subalgebras B, B', C, C', D such that $A = B \times C \times D = B' \times C' \times D$. Then $B \cong B'$ implies $C \cong C'$.

<u>Proof.</u> Let $B = B_1 \times \ldots \times B_n$ as a product of indecomposables. The result is proved by induction on n. If $n = 0$, $A = C \times D = C' \times D$, i.e. there are isomorphisms $f : C \times D \to A$, $g : A \to C' \times D$ with $f : (c,0) \mapsto c$ for $c \in C$, $g : c' \mapsto (c',0)$ for $c' \in C$, and $f : (0,d) \mapsto d$ and $g : d \mapsto (0,d)$ for $d \in D$. Let $j : C \to C \times D$; $c \mapsto (c,0)$ be the injection, and let $p : C' \times D \to C'$; $(c',d) \mapsto c'$ be the projection. Then $jfgp : C \to C'$ is an isomorphism, as required. So suppose the result is proved for the case where B is a product of less than n indecomposable factors. Since $B \cong B'$, $B' = B_1' \times \ldots \times B_n'$ for indecomposable B_i' isomorphic with B_i, $i = 1,\ldots,n$. Then $A = B_n \times (B_1 \times .. \times B_{n-1} \times C \times D) = B_1' \times .. \times B_n' \times C' \times D$. By the Pointed Exchange Theorem 422, either

(I) $\exists\, 1 \leqslant l \leqslant n$. $A = B_l' \times B_1 \times .. \times B_{n-1} \times C \times D$ or

(II) $C' = X \times Y$ and $A = B_1 \times .. \times B_{n-1} \times X \times C \times D$

$= B_1' \times .. \times B_n' \times B_n \times Y \times D$ or

(III) $D = X \times Y$ and $A = B_1 \times .. \times B_{n-1} \times X \times C \times D$

$= B_1' \times .. \times B_n' \times B_n \times C' \times Y$.

The three cases have to be examined separately.

Case (I):

$A = B_n \times (B_1 \times .. \times B_{n-1} \times C \times D) = B_l' \times (B_1 \times .. \times B_{n-1} \times C \times D)$.

By induction $B_n \cong B_l'$. Then $(B_1 \times .. \times B_{n-1}) \times B_n \cong (B_1' \times .. \times B_{l-1}' \times B_{l+1}' \times .. \times B_n') \times B_l'$. So by induction $B_1 \times .. \times B_{n-1} \cong B_1' \times .. \times B_{l-1}' \times B_{l+1}' \times .. \times B_n'$. Now $A =$

$(B_1 \times .. \times B_{n-1}) \times C \times (D \times B_l') = (B_1' \times .. \times B_{l-1}' \times B_{l+1}' \times .. \times B_n')$
$\times C' \times (D \times B_l')$. So by inductinn $C \cong C'$.

Case (II):

$A = (B_1 \times .. \times B_{n-1}) \times C \times (D \times B_n) = (B_1' \times .. \times B_{n-1}') \times (B_n' \times Y)$
$\times (D \times B_n)$, so by induction $C \cong B_n' \times Y$. Now $A =$
$(B_1 \times .. \times B_{n-1}) \times X \times (C \times D) = (B_1 \times .. \times B_{n-1}) \times B_n \times (C \times D)$, so by induction $B_n \cong X$. But $B_n' \cong B_n$. Thus $B_n' \cong X$, and
$C \cong X \times Y = C'$.

Case (III):

$A = (B_1 \times .. \times B_{n-1}) \times B_n \times (C \times D) = (B_1 \times .. \times B_{n-1}) \times X \times (C \times D)$.
So by induction, $X \cong B_n \cong B_n'$. Now $A =$
$(B_1 \times .. \times B_{n-1}) \times (X \times C) \times D = (B_1' \times .. \times B_{n-1}') \times (B_n' \times C') \times D$, so by induction $X \times C \cong B_n' \times C'$. Thus $B_n' \times C \cong B_n' \times C'$, i.e.
$\exists B'' , C'' \leqslant B_n' \times C' .\ B'' \times C'' = B_n' \times C'$, $B'' \cong B_n'$, $C \cong C''$. It remains to show $C'' \cong C'$. By the Pointed Exchange Theorem 422 either $B'' \times C'' = B_n' \times C'' = B_n' \times C' = B'' \times C'$, whence $C'' \cong C'$ by this theorem with $n = 0$, or else $C' = X' \times Y'$ and $B'' \times C'' = X' \times C''$
$= B'' \times Y' \times B_n' = B_n' \times C'$, whence $B'' \cong X'$ and $B_n' \times Y' \cong C''$ by this theorem with $n = 0$. Then $C' = X' \times Y' \cong B'' \times Y' \cong B_n' \times Y' \cong C''$, as required.]

With the help of the Pointed Cancellation Theorem at a

critical point, one can then prove the Cancellation Theorem showing that $[C] = [E]$ implies $C \simeq E$:

424 THEOREM For finite $\underline{\underline{T}}$-algebras B, C, D, E, $B \times C \cong D \times E$ and $B \cong D$ imply $C \simeq E$.

Proof. This done in three steps (a), (b), (c). All algebras appearing are finite.

(a) For Z_B, Z_D, Z_1, Z_2 in $\underline{\underline{Z}}$, $Z_B \times Z_1 \simeq Z_D \times Z_2$ and $Z_B \simeq Z_D$ imply $Z_1 \simeq Z_2$.

Proof of (a): By the Structure Theorem 418 for $\underline{\underline{Z}}$-algebras, there are $\underline{\underline{Z}}_0$-algebras Y_B, Y_D, Y_1, Y_2 such that $Z_B \simeq Y_B$, $Z_D \simeq Y_D$, $Z_1 \simeq Y_1$, $Z_2 \simeq Y_2$. Then $Z_B \times Z_1 \simeq Z_D \times Z_2$ implies $Y_B \times Y_1 \simeq Y_D \times Y_2$. By Corollary 419, $Y_B \times Y_1 \cong Y_D \times Y_2$ and $Y_B \cong Y_D$. By the Pointed Cancellation Theorem 423 $Y_1 \cong Y_2$. Hence $Z_1 \simeq Z_2$.

(b) For Z_1, Z_2 in $\underline{\underline{Z}}$, $B \times Z_1 \cong D \times Z_2$ and $B \cong D$ imply $Z_1 \simeq Z_2$.

Proof of (b): Since B and D are finite, they can be expressed as finite products of indecomposable factors. Each such factor is either in $\underline{\underline{Z}}$ or not in $\underline{\underline{Z}}$. Thus by collecting together these factors again, $B \cong B_n \times Z_B$ and $D \cong D_n \times Z_D$, where Z_B , Z_D are in $\underline{\underline{Z}}$ and B_n , D_n have no indecomposable factors in $\underline{\underline{Z}}$. Let $C \cong B \cong D$, $^2C = \beta'_n \sqcap \xi'_B = \delta'_n \sqcap \xi'_D$, with $\mathrm{Cnat}\overline{\beta'_n} \cong B_n$, $\mathrm{Cnat}\overline{\delta'_n} \cong D_n$, $\mathrm{Cnat}\overline{\xi'_B} \cong Z_B$, $\mathrm{Cnat}\overline{\xi'_D} \cong Z_D$.

Break up $\beta'_n \sqcap \xi'_B$ and $\delta'_n \sqcap \xi'_D$ into products of indecomposable congruences, and apply the Unique Factorisation Theorem 328. If an indecomposable factor β''_n of β'_n were centrally isomorphic with an indecomposable factor ξ''_D of ξ'_D , then by Proposition 415 $\mathrm{Cnat}\overline{\beta''_n}$ would be a central isotope of $\mathrm{Cnat}\overline{\xi''_D}$. But $\mathrm{Cnat}\overline{\xi''_D} \leqslant \mathrm{Cnat}\overline{\xi'_D} \in \underline{\underline{Z}}$, so by Proposition 412 $\mathrm{Cnat}\overline{\beta''_n}$, an indecomposable subalgebra of $\mathrm{Cnat}\overline{\beta'_n} \cong B_n$, would be in $\underline{\underline{Z}}$, contradictory to the choice of B_n . Thus the indecomposable factors of β'_n are centrally isomorphic with the factors of δ'_n , and those of ξ'_D with those of ξ'_B . Schematically, represent this as follows (as was done for the Unique Factorisation Theorem 328) :

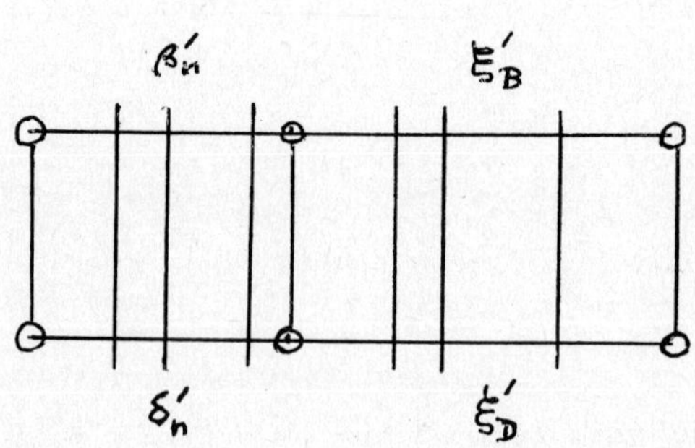

Then ξ'_B is centrally isomorphic with ξ'_D , whence by Proposition 415 $Z_B \cong \text{Cnat}\overline{\xi'_B} \simeq \text{Cnat}\overline{\xi'_D} \cong Z_D$.

Now let $A \cong B_n \times Z_B \times Z_1 \cong D_n \times Z_D \times Z_2$, where $^2A = \beta_n \sqcap \xi_B \sqcap \xi_1 = \delta_n \sqcap \xi_D \sqcap \xi_2$, $\text{Anat}\overline{\beta_n} \cong B_n$, $\text{Anat}\overline{\xi_B} \cong Z_B$, $\text{Anat}\overline{\xi_2} \cong Z_2$, etc. Break up $\beta_n \sqcap \xi_B \sqcap \xi_1$ and $\delta_n \sqcap \xi_D \sqcap \xi_2$ into indecomposables, apply the Unique Factorisation Theorem 328, and group similarly-behaved terms to get

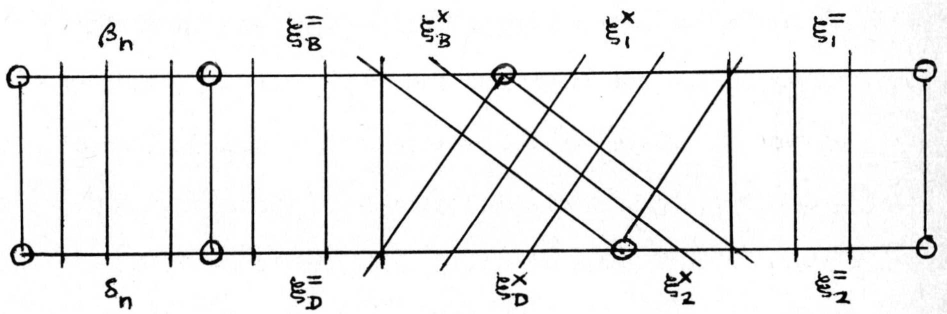

(again there are no cross-terms from β_n or δ_n) where $\xi_B = \xi_B^{=} \sqcap \xi_B^{x}$, etc., and ξ_B^x is centrally isomorphic with ξ_2^x , $\xi_B^{=}$ with $\xi_D^{=}$, etc. Since $\xi_B \sqcap \xi_1$ is centrally isomorphic with $\xi_D \sqcap \xi_2$, $Z_B \times Z_1 \simeq Z_D \times Z_2$. From above, $Z_B \simeq Z_D$. So by (a), $Z_1 \simeq Z_2$, as required for (b).

(c) $B \times C \cong D \times E$ and $B \cong D$ imply $C \simeq E$.

Proof of (c): by induction on the number of indecomposable factors of $B \times C \times D \times E$ not in $\underline{\underline{Z}}$. If C and E are in $\underline{\underline{Z}}$, (c) follows by (b). Otherwise, let $A \cong B \times C \cong D \times E$, ${}^2A = \beta \sqcap \gamma = \delta \sqcap \varepsilon$, $\mathrm{Anat}\overline{\beta} \cong B$, $\mathrm{Anat}\overline{\gamma} \cong C$, $\mathrm{Anat}\overline{\delta} \cong D$, $\mathrm{Anat}\overline{\varepsilon} \cong E$. Applying the Unique Factorisation Theorem 328 as before, one gets

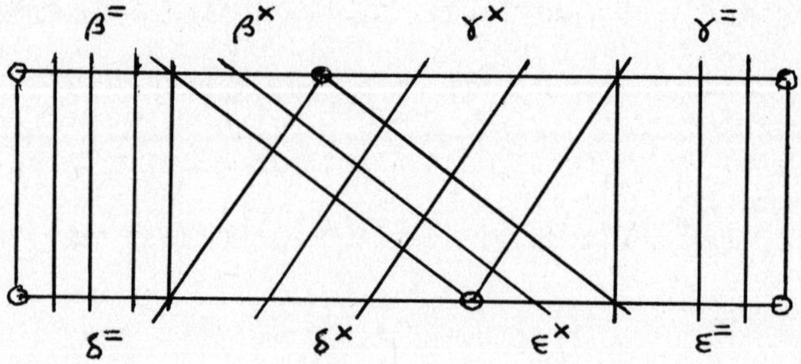

with $\beta = \beta^{=} \sqcap \beta^{x}$, etc., and $\beta^{=}$ centrally isomorphic with $\delta^{=}$, etc. By Proposition 415, $\mathrm{Anat}\overline{\beta^{=}} \simeq \mathrm{Anat}\overline{\delta^{=}}$. Then by Theorem 421, there is a finite $\underline{\underline{Z}}$-algebra Z such that $Z \times \mathrm{Anat}\overline{\beta^{=}} \cong Z \times \mathrm{Anat}\overline{\delta^{=}}$. Since $B \cong D$, $Z \times \mathrm{Anat}\overline{\beta^{=}} \times \mathrm{Anat}\overline{\beta^{x}} \cong Z \times B \cong Z \times D \cong Z \times \mathrm{Anat}\overline{\delta^{=}} \times \mathrm{Anat}\overline{\delta^{x}}$. Since $Z \times B \times Z \times D$ has fewer non-$\underline{\underline{Z}}$ indecomposable factors than $B \times C \times D \times E$, it follows by induction that $\mathrm{Anat}\beta^{x} \simeq \mathrm{Anat}\delta^{x}$. Then $\mathrm{Anat}\overline{\gamma^{x}} \simeq \mathrm{Anat}\overline{\delta^{x}} \simeq \mathrm{Anat}\overline{\beta^{x}} \simeq \mathrm{Anat}\overline{\varepsilon^{x}}$. Also $\mathrm{Anat}\overline{\gamma^{=}} \simeq \mathrm{Anat}\overline{\varepsilon^{=}}$. Thus $C \cong \mathrm{Anat}\overline{\gamma^{x}} \times \mathrm{Anat}\overline{\gamma^{=}} \simeq \mathrm{Anat}\overline{\varepsilon^{x}} \times \mathrm{Anat}\overline{\varepsilon^{=}} \cong E$. So $C \simeq E$.]

Thus for finite $\underline{\underline{T}}$-algebras B and C,

$$[B] = [C] \iff B \simeq C .$$

4.3 Cubic hypersurface quasigroups.

This section gives another example of the way in which central isotopy classes are more important than isomorphism classes. The example is drawn from a branch of quasigroup theory having applications to group theory and algebraic geometry. The basic reference for further information is Manin's book [Mi], also translated into English [MH].

A totally symmetric quasigroup or TS-quasigroup $(Q,.)$ is a quasigroup for which the ternary relation $\{ (x,y,z) \mid x.y = z \}$ on Q is invariant under all permutations of x , y , and z . A TS-quasigroup $(Q,.)$ is said to be Abelian (note the capital "A") [MH, 1.2] [Mi, I.1.2] iff for each element e in Q , $(Q,+,e)$ is an abelian group with identity e under the operation $x+y = u.(x.y)$. A cubic hypersurface quasigroup or CH-quasigroup (the name comes from geometrical applications) [MH, 1.3] [Mi, I.1.3] is a TS-quasigroup in which every set of three elements generates an Abelian subquasigroup. A commutative Moufang loop $(L,+,1)$ is a quasigroup $(L,+)$ having a nullary operation selecting the so-called identity element 1 and a unary operation assigning to each element x its so-called inverse $-x$, such that each set of two elements generates an abelian group and the <u>Moufang law</u> $x+(y+(x+z)) = (x+y+x)+z$ is satisfied. Manin gave the following theorem, which says roughly that CH-quasigroups are to commutative Moufang loops as Abelian quasigroups are to abelian groups:

431 THEOREM [MH, 5.1] [Mi, I.5.1] Let $(Q,.)$ be a CH-quasigroup, and e an arbitrary element of Q. Define the operation $+$ on Q by $x+y = e.(x.y)$, and let $F_e(Q,.) = (Q,+,e)$. Then $F_e(Q,.)$ is a commutative Moufang loop. If f is also an element of Q, the loops $F_e(Q,.)$ and $F_f(Q,.)$ are isomorphic.]

By Mal'cev's Theorem 142, the variety of commutative Moufang loops is a Mal'cev variety. The <u>centre</u> $Z(L,+,1)$ of a commutative Moufang loop $(L,+,1)$ is the equivalence class of 1 under the centre congruence $\zeta(L)$. It is left as an exercise for the reader to check that an element c of L is in $Z(L,+,1)$ iff $c+(x+y) = (c+x)+y$ for all x, y in L. Thus words in a commutative Moufang loop involving at most two elements not in the centre may be written as if they were words in an abelian group. Manin gave the following converse to the first part of Theorem 431 involving the notion of the centre of a commutative Moufang loop:

432 THEOREM [MH, 5.2] [Mi, I.5.2] Let $(L,+,1)$ be a commutative Moufang loop, and c an arbitrary element of its centre. Define the operation $.$ on L by $x.y = c - x - y$, and let $G_c(L,+,1) = (L,.)$. Then $(L,.)$ is a CH-quasigroup.]

Manin was unable, however, to give a converse to the second part of Theorem 431, for as he pointed out different choices of c in Theorem 432 may give non-isomorphic CH-quasigroups $G_c(L,+,1)$. In fact these different CH-quasigroups are central isotopes, and the various elements of the centre of the loop contrive to yield the full central isotopy class. More precisely :

433 THEOREM Let $(L,+,1)$ be a commutative Moufang loop, and let c, d be arbitrary elements of its centre. Then the CH-quasigroups $G_c(L,+,1)$ and $G_d(L,+,1)$ are centrally isotopic. Given a CH-quasigroup $(Q,.)$, all its central isotopes may be obtained as $G_dF_e(Q,.)$, where e is any chosen element of Q and d is in $ZF_e(Q,.)$.

Thus isomorphism classes of commutative Moufang loops

correspond exactly to

central isotopy classes of cubic hypersurface quasigroups.

The rest of this section is devoted to a proof of Theorem 433. The proof depends on a general proposition. For the moment, operator domains will be reintroduced to the notation, although actions will remain suppressed. One definition is needed : a <u>derived operation</u> on an Ω-algebra A is a mapping ${}^nA \to A$ of the form $(a_1,..,a_n) \mapsto (a_1,..,a_n,f_1,..,f_k)w$, where $(x_1,..,x_{n+k})w$ is an Ω-word in $\{x_1,..,x_{n+k}\}$, and f_1 , ... , f_k are certain fixed elements of A , for some natural number k .

434 PROPOSITION Let (A,Ω) be a Mal'cev algebra. Let Ω' be a set of derived operations obtained from Ω and including the Mal'cev operation, so that (A,Ω') is also a Mal'cev algebra. Then $(\zeta(A,\Omega),{}^2\Omega') \leqslant (\zeta(A,\Omega'),{}^2\Omega')$. If $({}^2A|\zeta(A,\Omega))$ is the congruence by which ${}^2(A,\Omega)$ centralises $(\zeta(A,\Omega),{}^2\Omega)$, then it is the congruence also by which ${}^2(A,\Omega')$ centralises $(\zeta(A,\Omega),{}^2\Omega')$.

<u>Proof.</u> Let $\omega' : (a_1,..,a_n) \mapsto (a_1,..,a_n,f_1,..,f_k)w$ be an element of Ω' . Suppose (a_1,a_1') , ... , $(a_n,a_n') \in \zeta(A,\Omega)$. By the reflexivity of $\zeta(A,\Omega)$, (f_1,f_1) , ... , $(f_k,f_k) \in \zeta(A,\Omega)$. Thus $(a_1..a_n\omega',a_1'..a_n'\omega') = ((a_1,..,a_n,f_1,..,f_k)w,(a_1',..,a_n',f_1,..,f_k)w) \in \zeta(A,\Omega)$. It follows that $\zeta(A,\Omega)$ is a congruence on (A,Ω') . If $({}^2A|\zeta(A,\Omega))$ is the congruence by which ${}^2(A,\Omega)$ centralises $(\zeta(A,\Omega),{}^2\Omega)$, a similar argument shows that $({}^2A|\zeta(A,\Omega))$ is a congruence on $(\zeta(A,\Omega),{}^2\Omega')$. Since the other conditions of Definition

211 are purely set-theoretic, it follows that $({}^2A|\zeta(A,\Omega))$ is a congruence by which ${}^2(A,\Omega')$ centralises $(\zeta(A,\Omega),{}^2\Omega')$. By Proposition 221, $({}^2A|\zeta(A,\Omega))$ is the unique congruence by which ${}^2(A,\Omega')$ centralises $(\zeta(A,\Omega),{}^2\Omega')$, and the very definition of $\zeta(A,\Omega')$ gives that $\zeta(A,\Omega) \subseteq \zeta(A,\Omega')$.]

Applying Proposition 434 to the contexts of Theorem 433,

435 $\zeta(Q,.) \subseteq \zeta F_e(Q,.) \subseteq \zeta G_{e.e}F_e(Q,.)$ and

436 $\zeta(L,+,1) \subseteq \zeta G_c(L,+,1) \subseteq \zeta F_1 G_c(L,+,1)$,

where $G_{e.e}$ is defined since $e.e \in ZF_e(Q,.)$ [MH, 5.1.3] [Mi, I.5.4]. Looking at 436, $1.(x.y) = c-(x.y) = c-(c-x-y) = x+y$, so that $F_1G_c(L,+,1) = (L,+,1)$. Also, by [MH, 5.1.2], [Mi, I.5.3], $G_{e.e}F_e(Q,.) = (Q,.)$. 435 and 436 thus yield :

437 PROPOSITION $\zeta(Q,.) = \zeta F_e(Q,.)$ and $\zeta(L,+,1) = \zeta G_c(L,+,1)$.]

Now the proof of Theorem 433 can be concluded. Firstly, let $(L,+,1)$ be a commutative Moufang loop, with c , d in $Z(L,+,1)$, so that $(c,d) \in \zeta(L,+,1) = \zeta G_c(L,+,1)$. Let $(L,.)$ be $G_c(L,+,1)$, i.e. $x.y = c-x-y$. Let $(L,*)$ be $G_d(L,+,1)$, i.e. $x*y = d-x-y$. Then $(x.y,x*y) = (c,d) - (x+y,x+y)$, so $(x.y,x*y)$ is in the coset $(c,d) + \hat{L}$ of $\hat{L}$ in $\zeta(L,+,1)$. But $(c,d) = (c,d) + (1,1)$ is also in this coset. It follows that $(x.y,x*y)$ $({}^2L|\zeta(L,+,1))$ (c,d), for the $({}^2L|\zeta(L,+,1))$-classes are just the cosets of $\hat{L}$ in $\zeta(L,+,1)$. By Proposition 434, $({}^2L|\zeta(L,+,1)) = ({}^2L|\zeta G_c(L,+,1))$. $G_d(L,+,1)$ is thus a central isotope of $G_c(L,+,1)$, as required.

For the second part of the theorem, let $(Q,*)$ be a central isotope of $(Q,.)$, say with $(e,f) \in \zeta(Q,.)$ and $\forall\, x, y \in Q$, $(x.y,x*y)$ $({}^2Q|\zeta(Q,.))$ (e,f). Define $(Q,+,e)$ to be $F_e(Q,.)$. By Proposition 437, $\zeta(Q,.) = \zeta(Q,+,e)$, and by Proposition 434 $({}^2Q|\zeta(Q,.))$ is a centreing congruence by which ${}^2(Q,+,e)$ centralises $\zeta(Q,+,e)$. Now

$$(e,f)\ ({}^2Q|\zeta(Q,.))\ (e.e,e.e) + (e,f) - (x+y,x+y)$$

$$= ((e.e)-(x+y),((e.e)+f)-(x+y))$$

$$= (x.y,((e.e)+f)-(x+y)) .$$

Let $d = (e.e)+f$. Note that d is in $ZF_e(Q,.)$. By hypothesis (e,f) $({}^2Q|\zeta(Q,.))$ $(x.y,x*y)$. It follows by (C1) that $x*y = d-x-y$, i.e. $(Q,*) = G_dF_e(Q,.)$, as required.

5 Plain algebras and equational completeness

O the pleasure of the plains.

G. F. HANDEL, Acis and Galatea.

Varieties of $\underline{\underline{T}}$-algebras are partially ordered by inclusion, and all varieties contain the trivial variety generated by the one-element $\underline{\underline{T}}$-algebra. The minimal varieties with respect to this partial ordering are of interest : they are described as *equationally complete*. An algebra generating such a variety and the set of identities satisfied by it are also called *equationally complete*.

A $\underline{\underline{T}}$-algebra A is said to be *simple* if the only congruences on it are $\hat{A}$ and ${}^2A$. A is said to be *plain* if it is finite, simple, and if its only subalgebras are singletons or all of A . Every variety containing a non-trivial finite algebra contains a plain algebra : take a minimal non-trivial quotient of a minimal non-trivial subalgebra of the non-trivial finite algebra. This chapter looks at equationally complete varieties, plain algebras, and the connections between them.

An equationally complete variety is generated by each non-trivial algebra in it, so every equationally complete variety containing a non-trivial finite algebra is generated by a plain algebra. Such varieties turn out to be characterised by these generating plain algebras, and consist entirely of algebras isomorphic to direct powers of the generator. Plain algebras which are not equationally complete are in the class $\underline{\underline{Z}}$, and the only non-trivial proper subvariety of the variety they generate is that generated by their central isotope with singleton subalgebra.

Throughout this chapter, $F_r(A)$ will denote the free algebra on r generators in the variety $\underline{\underline{T}}(A)$ generated by the $\underline{\underline{T}}$-algebra A . <u>Subquotient</u> will mean a quotient of a subalgebra.

5.1 Plain algebras.

511 PROPOSITION All subalgebras of a direct power of a plain algebra A are isomorphic to direct powers of A.

<u>Proof.</u> Let S be a subalgebra of ${}^nA$. For $i = 0, \ldots, n-1$, $\emptyset \neq S\pi^i \leqslant A$, and so $S\pi^i = A$ or $\{e_i\}$ for some singleton subalgebra $\{e_i\}$ of A. Discarding factors with $S\pi^i = \{e_i\}$, one may without loss of generality assume that $\forall\, i = 0, \ldots, n-1$, $S\pi^i = A$. Then since A is simple, each $\ker(\pi^i : S \to A)$ is a maximal congruence on S. Since $S \leqslant {}^nA$, $\bigcap_{i=0}^{n-1} \ker(\pi^i : S \to A) = \hat{S}$. Pick a minimal set of $\ker(\pi^i{:}S{\to}A)$ still having intersection $\hat{S}$, say without loss of generality $\{ \ker(\pi^i{:}S{\to}A) \mid i = 0,\ldots,m \}$. For each $i = 1,\ldots,m$, $\ker(\pi^i{:}S{\to}A) \leqslant \ker(\pi^i{:}S{\to}A) \circ \bigcap_{j=0}^{i-1} \ker(\pi^j{:}S{\to}A)$. If equality holds, $\bigcap_{j=0}^{i-1} \ker(\pi^j{:}S{\to}A) \leqslant \ker\pi^i$, so $(\bigcap_{j=0}^{i-1} \ker(\pi^j{:}S{\to}A)) \circ (\bigcap_{j=i+1}^{m} \ker(\pi^j{:}S{\to}A)) = \hat{S}$, contradicting the minimality of m. Thus for $i = 1,\ldots,m$, $\ker(\pi^i{:}S{\to}A) \circ \bigcap_{j=0}^{i-1} \ker(\pi^j{:}S{\to}A) = {}^2S$, whence $S \cong S\pi^0 \times \ldots \times S\pi^m = {}^{m+1}A$ as required.]

512 PROPOSITION If A is plain, $F_r(A)$ is isomorphic to a direct power of A.

<u>Proof.</u> The elements of $F_r(A)$ are words $(x_1,\ldots,x_r)w$ in the r generators $x_1, \ldots, x_r$ of $F_r(A)$. Define a morphism

$\theta : F_r(A) \to {}^{({}^nA)}A$; $(x_1,\ldots,x_r)w \mapsto ((a_1,\ldots,a_r) \mapsto (a_1,\ldots,a_r)w)$.
If $(x_1,\ldots,x_r)w\theta = (x_1,\ldots,x_r)w'\theta$, then $\forall\, a_1, \ldots, a_r \in A$, $(a_1,\ldots,a_r)w = (a_1,\ldots,a_r)w'$. Thus $(x_1,\ldots,x_r)w = (x_1,\ldots,x_r)w'$ is an identity of $\underline{\underline{T}}(A)$, i.e. $w = w'$ in $F_r(A)$. So θ embeds $F_r(A)$ as a subalgebra of ${}^{({}^rA)}A$. The result follows by Proposition 511.]

513 PROPOSITION If A is a plain algebra in $\underline{\underline{Z}}$, then ${}^2A/\hat{A}$ is also plain.

Proof. If S is a subalgebra of ${}^2A/\hat{A}$, let T be the union of the $({}^2A|{}^2A)$-classes forming S . Then T , as a subalgebra of ${}^2A$, has order $|A|$ or $|A|^2$ from Proposition 511. But each $({}^2A|{}^2A)$-class is of order $|A|$, and so S is of order 1 or $|A| = |{}^2A/\hat{A}|$.

Let U be a congruence on ${}^2A/\hat{A} = {}^2A\text{nat}({}^2A|{}^2A)$. Then U corresponds to a congruence U' on ${}^2A$ with $({}^2A|{}^2A) \leqslant U' \leqslant {}^4A$. But since U' is isomorphic to a direct power of A by Proposition 511, U' is $({}^2A|{}^2A)$ or ${}^4A$, whence U is $\widehat{{}^2A/\hat{A}}$ or ${}^2({}^2A/\hat{A})$. Thus ${}^2A/\hat{A}$ is plain.]

As a consequence of these properties, one may show that varieties generated by plain algebras are characterised by them :

514 THEOREM Let A , A' be plain algebras. Then $\underline{\underline{T}}(A) = \underline{\underline{T}}(A')$ iff $A \cong A'$.

<u>Proof.</u> Suppose $\underline{\underline{T}}(A) = \underline{\underline{T}}(A')$. Then $F_2(A) = F_2(A')$, and by Proposition 512 there are positive integers m , n such that ${}^{m}A \cong F_2(A) = F_2(A') \cong {}^{n}A'$. The diagonal $A\Delta_m$ is a subalgebra of ${}^{m}A$ isomorphic to A , and thus ${}^{n}A'$ has a corresponding subalgebra S isomorphic to A . By Proposition 511, S is isomorphic to a direct power of A' , and since S is simple this can only mean $A \cong S \cong A'$. The converse is trivial.]

5.2 <u>Equational completeness.</u>

As a consequence of Theorem 514, the equationally complete varieties containing non-trivial finite algebras are specified uniquely by equationally complete plain algebras. The next two theorems determine the latter, according to whether they have singleton subalgebras or not.

521 THEOREM If the plain algebra A has a singleton subalgebra $\{e\}$, then it is equationally complete.

<u>Proof.</u> It must be shown that A is in the variety generated by each non-trivial subquotient of a direct power of A . By Proposition 511 subalgebras of direct powers are themselves direct powers, and so it suffices to show that A is in $\underline{\underline{T}}(A')$ for each non-trivial quotient A' of a direct power of A .

Let V be a proper congruence on ^nA with $A' \cong {}^nA\mathrm{nat}V$, and assume without loss of generality that n is minimal with respect to the property that A' is a quotient of ^nA . Let $B = \{ x \in {}^nA \mid \forall\, i > 0 ,\ x\pi^i = e \}$, and let $W = V \cap {}^2B$. Then $W = \hat{B}$ or ^2B . Suppose $W = {}^2B$. Then for all $(a_1,a_2,\dots,a_n) \in {}^nA$,

$(a_0,a_1,\dots,a_{n-1})\ V\ (a_0,a_1,\dots,a_{n-1})$,

$(e\ ,e\ ,\dots,\ e\ )\ V\ (a_0,e\ ,\dots,\ e\ )$ since $W = {}^2B$,

and $(e\ ,e\ ,\dots,\ e\ )\ V\ (e\ ,e\ ,\dots,\ e\ )$.

$\Rightarrow$ $(a_0,a_1,\dots,a_{n-1})\ V\ (e\ ,a_1,\dots,a_{n-1})$, applying the Mal'cev operation P . Hence $A' = (\pi^0)^{-1}(e)\mathrm{nat}(V \cap {}^2(\pi^0)^{-1}(e))$, a quotient of $(\pi^0)^{-1}(e) = {}^{n-1}A$. But this contradicts the minimality of n . Thus $W = \hat{B}$, and $A \cong B \cong B\mathrm{nat}V$, which is isomorphic to a subalgebra of A' .]

522 THEOREM A plain algebra A without singleton subalgebras is equationally complete iff it is not in $\underline{\underline{Z}}$.

<u>Proof.</u> Let A be in $\underline{\underline{Z}}$. By Proposition 513, $^2A/\hat{A}$ is plain.

Since A has no singleton subalgebras, $A \not\cong {}^2A/\hat{A}$, and so by Theorem 514, $\underline{\underline{T}}({}^2A/\hat{A})$ is a proper non-trivial subvariety of $\underline{\underline{T}}(A)$. Thus A is not equationally complete.

Conversely, suppose A is not equationally complete : let $\underline{\underline{T}}(A)$ have the non-trivial proper subvariety $\underline{\underline{T}}'$. By Proposition 512, $F_2(A) \cong {}^nA$ for some positive integer n . Then $F_2(\underline{\underline{T}}')$, the free $\underline{\underline{T}}'$-algebra on two generators, satisfies some identity I not satisfied by A , and is isomorphic to the quotient of ${}^nA$ by some proper congruence V . Fix a in A , and let $B = < a\Delta_n \mathrm{nat} V >$. B , as a subalgebra of ${}^nA \mathrm{nat} V$, satisfies I , but it is also the image of A under $\Delta_n \mathrm{nat} V$, and so must be trivial. Thus $A\Delta_n$ is contained within a single V-class.

Let m be minimal with respect to the property that there is a proper congruence V on ${}^mA$ such that $A\Delta_m$ is contained within a single V-class. Certainly $m \geqslant 2$. If $m = 2$, plainness of A and properness of V imply that $\hat{A}$ is a V-class, and so by Proposition 223 A is in $\underline{\underline{Z}}$.

So suppose $m > 2$. Let $S = \{ x \in {}^mA \mid x\pi^0 = x\pi^1 \}$. $S \cong {}^{m-1}A$, and $A\Delta_m$ is contained within a single $V \cap {}^2S$-class . Minimality of m then implies that $S \mathrm{nat} V$ is trivial, whence

$S^V = A\Delta_m{}^V$. Thus $|A|^{m-1} \leqslant |A\Delta_m{}^V| < |A|^m$, from which it follows that $A\Delta_m{}^V = S$. Define a congruence U on ${}^2A$ by $(x,y)\ U\ (x',y')$ $\Leftrightarrow \exists\ a_2, \ldots, a_{m-1} \in A$. $(x,y,a_2,\ldots,a_{m-1})\ V\ (x',y',a_2,\ldots,a_{m-1})$. Then $\hat{A}$ is a U-class. But this contradicts the minimality of m .]

Letting Plain denote the class of plain $\underline{\underline{T}}$-algebras, and EC the class of equationally complete $\underline{\underline{T}}$-algebras, one may summarise the situation schematically as:

$$\underline{\text{Plain}} - \underline{\text{EC}} = \underline{\text{Plain}} \cap (\underline{\underline{Z}}(\underline{\underline{T}}) - \underline{\underline{T}}_0) .$$

If A is a plain algebra that is not equationally complete, one can still say a lot about the variety generated by it :

523 THEOREM Let A be a plain $\underline{\underline{Z}}(\underline{\underline{T}})$-algebra without singleton subalgebras. Then:

(i) $\underline{\underline{T}}({}^2A/\hat{A})$ is the only non-trivial proper subvariety of $\underline{\underline{T}}(A)$.

(ii) $F_r(A) \cong {}^nA \Rightarrow F_r({}^2A/\hat{A}) \cong {}^{n-1}({}^2A/\hat{A})$.

(iii) If S is the collection of identities satisfied by A , then there is an identity I in one variable such that $S \cup \{I\}$ is equationally complete.

__Proof.__ Let $V_i = \ker(f : {}^iA \to {}^{i-1}({}^2A/\hat{A}) \; ; \; (a_0,\dots,a_{i-1}) \mapsto$

$$((a_0,a_1)^{({}^2A|{}^2A)},(a_0,a_2)^{({}^2A|{}^2A)},\dots,(a_0,a_{i-1})^{({}^2A|{}^2A)}))$$

for $i \geqslant 2$. Let A' in $\underline{\underline{T}}(A)$ satisfy an identity I not satisfied by A . A' is isomorphic with the quotient of some power ${}^lA$ of A by a congruence V . $A\Delta_l \mathrm{nat} V$ must be trivial or isomorphic with A , since A is plain, and as a subalgebra of A' satisfying I it cannot be isomorphic with A . Thus $A\Delta_l$ is contained within a single V-class, whence $V_l \leqslant V$. Thus A' is a quotient of ${}^lA\mathrm{nat}V_l$ $\cong {}^{l-1}({}^2A/\hat{A})$, and thus is in $\underline{\underline{T}}({}^2A/\hat{A})$.

Suppose $F_r(A) \cong {}^nA$. Then $F_r({}^2A/\hat{A})$, satisfying identities not satisfied by A , is isomorphic with a proper quotient of ${}^nA$, say by congruence V . As above, $V_n \leqslant V$. Thus $F_r({}^2A/\hat{A})$ is isomorphic with a quotient of ${}^nA\mathrm{nat}V_n \cong {}^{n-1}({}^2A/\hat{A})$. But ${}^nA\mathrm{nat}V_n$ is an r-generator algebra in $\underline{\underline{T}}({}^2A/\hat{A})$, and thus is a morphic image of $F_r({}^2A/\hat{A})$. Hence $F_r({}^2A/\hat{A}) = {}^{n-1}({}^2A/\hat{A})$.

Let J be an identity satisfied by ${}^2A/\hat{A}$ but not by A . Since $\underline{\underline{T}}({}^2A/\hat{A})$ is the only non-trivial proper subvariety of $\underline{\underline{T}}(A)$, $S \cup \{J\}$ is equationally complete. Since $F_1({}^2A/\hat{A})$ is a proper image of $F_1(A)$, it satisfies an identity I in one variable not satisfied by A . Then $S \cup \{I\}$ is equationally complete.]

The structure of finite algebras in an equationally complete variety is easily described by the following :

524 THEOREM Finite members of the variety $\underline{\underline{T}}(A)$ generated by an equationally complete plain algebra A are isomorphic to direct powers of A .

<u>Proof.</u> By Proposition 511 it suffices to show that quotients of direct powers of A are themselves isomorphic to powers of A . Certainly the only quotients of ${}^{1}A$, namely ${}^{1}A$ and ${}^{0}A$, are direct powers of A . Assume as an induction hypothesis that for all positive integers m less than a positive integer n greater than 1 , all quotients of ${}^{m}A$ are powers of A . Let V be a congruence on ${}^{n}A$. If $V = \widehat{{}^{n}A}$, ${}^{n}A\text{nat}V \cong {}^{n}A$, so assume $\widehat{{}^{n}A} < V$. Let $A_{ij} = \{ x \in {}^{n}A \mid x\pi^{i} = x\pi^{j} \}$, $S_{ij} = \bigcup A_{ij}{}^{V}$, $i \neq j$. If $S_{ij} = {}^{n}A$, ${}^{n}A\text{nat}V = S_{ij}\text{nat}V = A_{ij}\text{nat}V$ is a morphic image of ${}^{n-1}A$, and so by induction a direct power of A .

Suppose $S_{ij} < {}^{n}A$ for all $i \neq j$. Then $A_{ij} \leqslant S_{ij}$ and $|A_{ij}| = |A|^{n-1}$ imply $A_{ij} = S_{ij}$, so that each A_{ij} is a union of V-classes. Fix $a \in A$. Then $a\Delta_n \in A_{ij}$ for all $i \neq j$. If $x \; V \; a\Delta_n$, then $x \in A_{ij}$ for all $i \neq j$, whence $x \in A\Delta_n$. Thus $a\Delta_n{}^{V}$ is a subset of the diagonal. Since V induces a congruence on the simple $A\Delta_n$, and $V > \widehat{A}$, $A\Delta_n$ is a V-class. Let $B =$

$\{ (a_0,a_1,\ldots,a_{n-1}) \in {}^nA \mid a_1 = a_2 = \ldots = a_{n-1} \}$. Now $\theta : {}^2A \to B$; $(a_0,a_1) \mapsto (a_0,a_1,\ldots,a_1)$ is an isomorphism, and $V \cap {}^2B$ has $A\theta$ as a congruence class. Thus by Proposition 223, $A \in \underline{\underline{Z}}$, and since A is equationally complete it must have a singleton subalgebra, so that $A \cong {}^2A/\hat{A}$. Then ${}^nA\mathrm{nat}V = {}^nA/(AA_n) \cong {}^{n-1}({}^2A/\hat{A}) \cong {}^{n-1}A$, as required.]

A similar argument but with a different induction hypothesis yields the following :

525 THEOREM Finite members of the variety $\underline{\underline{T}}(A)$ generated by a plain algebra A that is not equationally complete are isomorphic to direct products of powers of A with powers of ${}^2A/\hat{A}$.]

5.3 Free algebras in equationally complete varieties.

Section 5.1 showed that the free algebra in an equationally complete variety (containing a non-trivial finite algebra) on a finite number m of generators is some direct power, say the d(m)-th, of the unique equationally complete plain algebra A within the variety. The function d from the positive integers to themselves is a measure of the complexity of the variety. This section sets out to specify the function d in terms of invariants of A : its size $|A|$, the number $|\{\{e\} \leqslant A\}|$ of singleton subalgebras it has, and the order of its automorphism group $\mathrm{Aut}A$. An interesting dichotomy takes place. If A is not in $\underline{\underline{Z}}$, then d is asymptotically exponential in m (Theorem 535), while if A is in $\underline{\underline{Z}}$ (and thus has a singleton subalgebra), d has a linear upper bound (Theorem 536).

Firstly, until the end of Theorem 535, consider the case that A is a plain algebra not in $\underline{\underline{Z}}$ (and so equationally complete). Fix m . Let $\underline{x}$ denote the m-tuple $(x_1,..,x_m)$ of elements of A . If $\{e\}$ is a singleton subalgebra of A , let $\underline{e} = (e,..,e)$ in ${}^mA$. Let the automorphism group $\mathrm{Aut}A$ of A have the action $\alpha : (x_1,..,x_m) \mapsto (x_1\alpha,..,x_m\alpha)$ on ${}^mA$. A non-empty subset $\{\underline{x}_1,..,\underline{x}_s\}$ of ${}^mA$ is called an <u>independent representing subset</u> or IRS if and only if the subalgebra $< (x_{1i},..,x_{si}) \mid i = 1,..,m >$ of ${}^sA$ generated by $\{ (x_{1i},..,x_{si}) \mid i = 1,..,m \}$ is isomorphic to ${}^sA$. Note that for a singleton subalgebra $\{e\}$ of A , $\underline{e}$ cannot be a member of an IRS. If $\{\underline{x}_1,..,\underline{x}_s\}$ is an IRS of maximal size s , and $\underline{x}$ is an element of ${}^mA$, then for all w , w' in $F_m(A)$, $(x_1,..,x_m)w = (x_1,..,x_m)w'$ if and only if $(x_{j1},..,x_{jm})w = (x_{j1},..,x_{jm})w'$ for $j = 1,..,s$, lest $< (x_{1i},..,x_{si},x_i) \mid i = 1,..,m >$ be isomorphic to ${}^{s+1}A$. Thus $d(m) = s$.

531 PROPOSITION Let A be a plain algebra not in $\underline{\underline{Z}}$. Let $\{\underline{y}_1,..,\underline{y}_s\}$ be an IRS. If $\underline{y}$ is an element of A , then precisely one of the following two statements holds :

(i) $\underline{y}$ is not in $\{\underline{y}_1,..,\underline{y}_s\}$ and $\{\underline{y}_1,..,\underline{y}_s,\underline{y}\}$ is an IRS ;

(ii) the assignments $(y_{1i},..,y_{si}) \to (y_{1i},..,y_{si},y_i)$ for $i = 1,..,m$ generate an isomorphism $\varphi(\underline{y}_1,..,\underline{y}_s;\underline{y})$ from $< (y_{1i},..,y_{si}) \mid i = 1,..,m >$ to $< (y_{1i},..,y_{si},y_i) \mid i = 1,..,m >$.

Proof. $< (y_{1i},..,y_{si},y_i) \mid i = 1,..,m >$ is isomorphic either to ${}^{s+1}A$ or to ${}^{s}A$. In the former case, (i) holds, and in the latter case, (ii) holds.]

532 PROPOSITION Let A be a plain algebra not in $\underline{\underline{Z}}$. Let R be a subset of ${}^{m}A$, and let $\underline{a}$, $\underline{b}$ be two distinct elements of ${}^{m}A$ such that $R \cup \{\underline{a}\}$ and $R \cup \{\underline{b}\}$ are independent representing subsets. Then precisely one of the following must be true :

(i) $R \cup \{\underline{a},\underline{b}\}$ is an IRS ;

(ii) there is an automorphism α of A such that $\underline{a}\alpha = \underline{b}$.

Proof. Firstly, suppose R empty. $\{\underline{a}\}$ and $\{\underline{b}\}$ are independent representing subsets. If $\{\underline{a},\underline{b}\}$ is not an IRS, then by option (ii) of Proposition 531 with $s = 1$, $\underline{y}_1 = \underline{a}$, $\underline{y} = \underline{b}$, and again with $s = 1$, $\underline{y}_1 = \underline{b}$, $\underline{y} = \underline{a}$, $\varphi(\underline{a};\underline{b})\varphi(\underline{b};\underline{a})^{-1}$ is an automorphism of A mapping $\underline{a}$ to $\underline{b}$.

Now suppose that $R = \{\underline{x}_1,..,\underline{x}_r\}$ with $r \geq 1$, and that $R \cup \{\underline{a},\underline{b}\}$ is not an IRS. It will be shown that either option (ii) of the proposition holds, or else there is the contradiction that A is in $\underline{\underline{Z}}$. Let $S = \langle (x_{1i},..,x_{ri},a_i,b_i) \mid i = 1,..,m \rangle$. By Proposition 531, there are isomorphisms

$\varphi(\underline{x}_1,..,\underline{x}_r,\underline{a};\underline{b}) : {}^{r+1}A \to S \; ; \; (x_{1i},..,x_{ri},a_i) \mapsto (x_{1i},..,x_{ri},a_i,b_i)$ and

$\varphi(\underline{x}_1,..,\underline{x}_r,\underline{b};\underline{a}) : {}^{r+1}A \to S \; ; \; (x_{1i},..,x_{ri},b_i) \mapsto (x_{1i},..,x_{ri},a_i,b_i)$

(note the slight twist from the notation of Proposition 531 at the end of the latter). Define projections $\pi_{ij..k} : {}^nA \to A \times A \times .. \times A$; $(z_1,..,z_n) \mapsto (z_i,z_j,..,z_k)$. Let $T = \ker(\pi_{12..r} : S \to {}^rA)$. Note that $T(\pi_1\pi_{r+2},\pi_2\pi_{r+2}) = {}^2A$. Let $U = \ker((\pi_1\pi_{r+1},\pi_2\pi_{r+1}) : T \to {}^2A)$. If $(((p_1,..,p_r,x,y),(p_1,..,p_r,x',y')),((q_1,..,q_r,x,z),(q_1,..,q_r,x',z')))$ is in U , then $y = y'$

$$\Rightarrow x = (p_1,..,p_r,y)\varphi(\underline{x}_1,..,\underline{x}_r,\underline{b};\underline{a})\pi_{r+1}$$

$$= (p_1,..,p_r,y')\varphi(\underline{x}_1,..,\underline{x}_r,\underline{b};\underline{a})\pi_{r+1} = x'$$

$$\Rightarrow z = (q_1,..,q_r,x)\varphi(\underline{x}_1,..,\underline{x}_r,\underline{a};\underline{b})\pi_{r+2}$$

$$= (q_1,..,q_r,x')\varphi(\underline{x}_1,..,\underline{x}_r,\underline{a};\underline{b})\pi_{r+2} = z'$$. Let $V =$ $U((\pi_1\pi_1\pi_{r+2},\pi_1\pi_2\pi_{r+2}),(\pi_2\pi_1\pi_{r+2},\pi_2\pi_2\pi_{r+2}))$. Then for $((y,y'),(z,z'))$ in V ,

533 $\quad y = y' \Rightarrow z = z'$.

Since $T(\pi_1\pi_{r+2},\pi_2\pi_{r+2}) = {}^2A$, $\widehat{{}^2A} \leq V$. Thus V is a congruence on ${}^2A$. Let $W = (U \cap {}^2\hat{S})(\pi_1\pi_1\pi_{r+2},\pi_2\pi_1\pi_{r+2}) = \{ (y,z) \mid$ $\exists\, (((p_1,..,p_r,x,y),(p_1,..,p_r,x,y)),((q_1,..,q_r,x,z),(q_1,..,q_r,x,z))) \in U \}$.

$\hat{A} \leqslant W \leqslant {}^2A$, so W is a congruence on A . Since A is simple, $W = A$ or $W = {}^2A$. If $W = A$, $S\pi_{(r+1)(r+2)}$ is an automorphism of A mapping $\underline{a}$ to $\underline{b}$, and option (ii) of the proposition holds. If $W = {}^2A$, $\hat{A}$ is contained in a congruence class for V , and 533 shows that this congruence class contains nothing but elements of $\hat{A}$. Thus A is in $\underline{\underline{Z}}$, a contradiction.]

534 PROPOSITION Let A be a plain algebra not in $\underline{\underline{Z}}$. A non-empty subset $\{\underline{x}_1,\ldots,\underline{x}_s\}$ of ${}^mA - \{\ \underline{e} \mid \{e\} \leqslant A\ \}$ is an IRS if and only if the $\underline{x}_i$ are in distinct (AutA)-orbits.

Proof. If $\underline{x}\alpha = \underline{y}$ for some α in AutA , $< (x_1,\ldots,x_m),(y_1,\ldots,y_m) > \cong A$, so that $\{\underline{x},\underline{y}\}$ cannot be a subset of an IRS. Conversely, it will be shown by induction on s that if $\underline{x}_1$, ... , $\underline{x}_s$ are in distinct (AutA)-orbits, $\{\underline{x}_1,\ldots,\underline{x}_s\}$ is an IRS. If $s = 1$, the result is trivial. If $s = 2$, it follows by option (ii) of Proposition 532 with R empty, $\underline{a} = \underline{x}_1$, $\underline{b} = \underline{x}_2$. Otherwise, let $R = \{\underline{x}_1,\ldots,\underline{x}_{s-2}\}$, $\underline{a} = \underline{x}_{s-1}$, $\underline{b} = \underline{x}_s$. By induction, $R \cup \{\underline{a}\}$ and $R \cup \{\underline{b}\}$ are independent representing subsets. Since $\underline{a}$ and $\underline{b}$ are in distinct (AutA)-orbits, they are distinct, so Proposition 532 applies, and its option (i) must hold : $R \cup \{\underline{a},\underline{b}\} = \{\underline{x}_1,\ldots,\underline{x}_s\}$ is an IRS.]

535 THEOREM Let A be a plain algebra not in $\underline{\underline{Z}}$. Then for each

positive integer m , the free algebra $F_m(A)$ on m generators in the variety generated by A is isomorphic to ${}^{d(m)}A$, where $d(m) = (|A|^m - |\{\{e\} \leqslant A\}|) / |\mathrm{Aut}A|$.

<u>Proof.</u> Let $\underline{x}$ be in ${}^{m}A - \{\underline{e} | \{e\} \leqslant A\}$. Since A is plain, every element of A can be expressed as a word $(x_1,..,x_m)w$ in the components of $\underline{x}$ for some m-ary operation w in $F_m(A)$. If α in AutA fixes $\underline{x}$, then $(x_1,..,x_m)w\alpha = (x_1\alpha,..,x_m\alpha)w = (x_1,..,x_m)w$ for all w in $F_m(A)$, so that α is the identity automorphism. Thus the orbit of $\underline{x}$ under AutA has |AutA| elements. By Proposition 534 d(m) , the maximal size of an IRS, is the number of orbits of AutA on ${}^{m}A - \{\underline{e} | \{e\} \leqslant A\}$. Since each orbit has |AutA| elements, this number is as stated.]

Thus for equationally complete varieties not within $\underline{\underline{Z}}$, the function d is asymptotically exponential.

Now consider the case that A is an equationally complete plain algebra in $\underline{\underline{Z}}$, thus having a singleton subalgebra {0} . A may be regarded as a $\underline{\underline{Z}}_0$-algebra, its element 0 being selected by a nullary operation. Theorem 536 below gives the upper bound $m.\log_{(|\mathrm{Aut}A|/|\{\{e\} \leqslant A\}|)+1} |A|$ for d(m) in this case. The upper bound is attained if the nullary operation selecting 0 is an operation of $\underline{\underline{T}}$, but may not be attained otherwise. For example, consider the quasigroup (A,.) with multiplication table

.	0	1	2
0	0	2	1
1	2	1	0
2	1	0	2

.

AutA is the symmetric group on $\{0,1,2\}$ of order 6 , and there are 3 idempotents, so the upper bound given for d(1) is 1 . But d(1) is 0 , since all the elements of A are idempotent. Now suppose that selection of 0 is a nullary operation of A (so A is being regarded as a pointed quasigroup, in the sense of Section 1.2). Then AutA just consists of the identity and the transposition exchangeing 1 and 2 , while $\{0\}$ is the only singleton subalgebra, so again the upper bound given for d(1) is 1 . This time however the bound is attained, since there are now non-trivial unary operations, e.g. multiplication by 0 .

536 THEOREM Let A be an equationally complete plain algebra in $\underline{\underline{Z}}$, thus having a singleton subalgebra $\{0\}$. Then for each positive integer m , the free algebra $F_m(A)$ on m generators in the variety generated by A is isomorphic to ${}^{d(m)}A$, where

$$d(m) \leqslant m.\log_{(|\mathrm{Aut}A|/|\{\{e\} \leqslant A\}|)+1} |A| \quad .$$

<u>Proof.</u> By the Structure Theorem 418 for $\underline{\underline{Z}}$-algebras, A as a $\underline{\underline{Z}}_0$-algebra is an abelian group in the category of $\underline{\underline{T}}$-algebras, as described in Section 1.2. There is a binary operation $+$ on A making A an abelian group with identity 0, and such that $+ : {}^2A \to A$ is a morphism of $\underline{\underline{T}}$-algebras. If w is in $F_m(A)$, define for $i = 1,..,m$ the mapping $t_{iw} : A \to A ; x \mapsto (0,..,x,..,0)w$, where the x appears as the i-th argument of w. Then

$$\begin{aligned}(x_1,..,x_m)w &= (x_1+0,0+x_2,..,0+x_m)w\\ &= (x_1,0,..,0)w + (0,x_2,..,x_m)w\\ &= \dots\\ &= (x_1,0,..,0)w + (0,x_2,..,0)w + (0,0,..,x_m)w\\ &= x_1t_{1w} + x_2t_{2w} + \dots + x_mt_{mw} .\end{aligned}$$

Let T be the set of unary operations of the variety generated by A in $\underline{\underline{Z}}_0$. Then the mapping $F_m(A) \to {}^mT ; w \mapsto (t_{1w},t_{2w},..,t_{mw})$ is a monomorphism of $\underline{\underline{T}}$-algebras. (It is an isomorphism if the nullary operation selecting 0 is in $F_0(A)$.) The proof reduces to showing that the size of T is $|A|$ raised to the power

$\log_{(|\mathrm{Aut}A|/|\{\{e\} < A\}|)+1}|A|$.

T has an addition $+$ defined by $t+t' : x \mapsto xt + xt'$, and a multiplication defined by composition. It becomes a ring, and A becomes a T-module. Sub- $\underline{\underline{Z}}_0$ -algebras of A correspond to sub-T-modules, so A , which is plain as a $\underline{\underline{T}}$-algebra, is irreducible as a T-module. By the definition of T, A is a faithful T-module, so T is a primitive ring. Let E be the ring of T-endomorphisms of the T-module A. Note that A, T, and E are all finite. Then T is the full ring of E-endomorphisms of A regarded as a module over the division ring E [Ja, Theorem III.2.1]. Since E is finite, it is a field, by the Wedderburn Theorem [Ja, Theorem VII.12.1] Let its order be p^b for some prime p and positive integer b. A is an E-vector space, say of dimension a. Then the order of A is p^{ab} and the order of T, the ring of a x a matrices over E, is p^{a^2b}.

E*, the multiplicative group of non-zero elements of E, is the group of T-automorphisms of the T-module A. It consists of the stabiliser of 0 in the group of automorphisms of the $\underline{\underline{T}}$-algebra A. The complement of this stabiliser is an abelian group of fixed-point-free automorphisms isomorphic to the subgroup $\{ e \mid \{e\} \leqslant A \}$ of the abelian group (A,+,0) (see 414). Thus $|E^*| = |\mathrm{Aut}A|/|\{\{e\} \leqslant A\}| = p^b - 1$. Now $|T| = |A|^a$, and $|A| = (p^b)^a$. So $a = \log_{(|\mathrm{Aut}A|/|\{\{e\} \leqslant A\}|)+1} |A|$, as required.]

NOTES

The results of the first two sections of this chapter are basically due to Bruce Caine [Ca], who proved them initially for quasigroups as part of his doctoral dissertation under Trevor Evans' supervision at Emory University. The argument of Proposition 511 is due to Foster and Pixley [FP, Theorem 2.4]. Plain quasigroups were first studied by Helen Popova-Alderson, in a series of papers dating back to the early fifties (see the references in [PA]). Note however that she excluded quasigroups with idempotents (singleton subalgebras) from her definition. Perhaps her original French term <u>uni</u> should be used for this. In [PA] she proved Theorem 535 for the case of quasigroups with $m = 1$ and $|\{\{e\} \leq A\}| = 0$, calling $F_1(A)$ the <u>logarithmetic</u> of A. In broad outline the proof of Theorem 535 above is a generalisation of her method. Lemma 2 of [PA], however, the analogue of Proposition 532, makes no distinction between $\underline{\underline{Z}}$ and non-$\underline{\underline{Z}}$ quasigroups, and is thus false as it stands (the error in the proof in [PA], noticed by Caine, being the statement in line 11 of page 5 that " $M(+)$ is a quasigroup").

6 Extensions and obstructions

> Two competent men, provided with the necessary hand signals and detonators, must be appointed to protect the obstruction, one on each side.
>
> BRITISH RAILWAYS, Rule Book 1950.

Chapters 2 and 3 looked at algebras under a given $\underline{\underline{T}}$-algebra, i.e. its quotients. This chapter considers algebras over a given $\underline{\underline{T}}$-algebra, i.e. its extensions. It asks whether extensions satisfying certain properties exist, classifies them if they do, and considers the obstruction to their existence if they don't. The subject requires a modicum of homological machinery which is introduced in the first section.

6.1 Simplicial objects and cohomology.

Let ε_n^i be the operation which deletes the (i+1)-th of any list of n symbols: $\varepsilon_n^i : 01\ldots(i-1)i(i+1)\ldots(n-1) \mapsto 01\ldots(i-1)(i+1)\ldots(n-1)$. Let δ_n^i be the operation which repeats the (i+1)-th of a list of n symbols: $\delta_n^i : 01\ldots(i-1)i(i+1)\ldots(n-1) \mapsto 01\ldots(i-1)ii(i+1)\ldots(n-1)$. These operations for all natural numbers n and $i < n$ generate (the morphisms of) a category $\mathbb{\Delta}$ called the <u>simplicial category</u>. A <u>simplicial object</u> in $\underline{\underline{T}}$ is (the image of) a functor $\mathbb{\Delta} \to \underline{\underline{T}}$. A <u>simplicial map</u> is a natural transformation between the functors. Thus a simplicial object B^* consists of a $\underline{\underline{T}}$-algebra B^n for each natural number n and morphisms between these (usually labelled by their preimages in $\mathbb{\Delta}$):

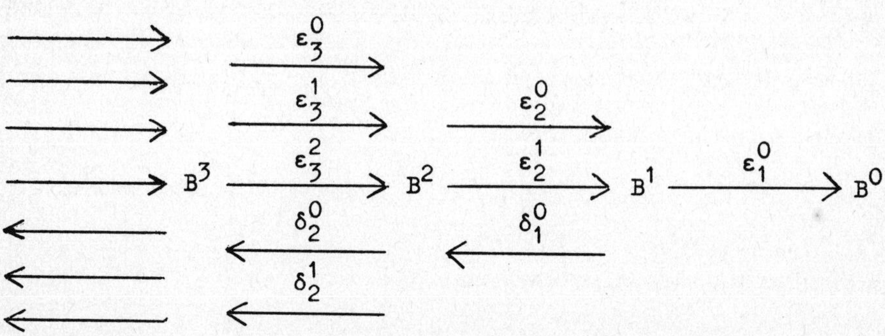

A simplicial map $p^* : B^* \to C^*$ consists of $\underline{\underline{T}}$-morphisms $p^n : B^n \to C^n$ for each natural number n such that all the diagrams

$$\begin{array}{ccc} B^n & \xrightarrow{\varepsilon_n^i} & B^{n-1} \\ {\scriptstyle p^n}\downarrow & & \downarrow{\scriptstyle p^{n-1}} \\ C^n & \xrightarrow{\varepsilon_n^i} & C^{n-1} \end{array} \quad \text{and} \quad \begin{array}{ccc} B^{n+1} & \xrightarrow{\delta_n^i} & B^n \\ {\scriptstyle p^{n+1}}\downarrow & & \downarrow{\scriptstyle p^n} \\ C^{n+1} & \xrightarrow{\delta_n^i} & C^n \end{array} \quad \text{commute.}$$

The suffices n of ε_n^i and δ_n^i are usually dropped, and $\varepsilon_1^0 : B^1 \to B^0$, $\delta_1^0 : B^1 \to B^2$ respectively are usually just called ε and δ . Sometimes the δ_j^i are not explicitly mentioned.

If $Z \to B^0$ is a module over B^0 , i.e. an abelian group in the category $\underline{\underline{T}}/B^0$, one may form pullbacks

$$\begin{array}{ccc} Z^n & \longrightarrow & Z \\ \downarrow & & \downarrow \\ B^n & \xrightarrow{\varepsilon_n^0 \varepsilon_{n-1}^0 \cdots \varepsilon_1^0} & B^0 \end{array}$$

for each natural number n . This produces modules Z^n over B^n . One speaks of a <u>module over the simplicial object</u> B^* , and denotes the B^n-module $Z^n \to B^n$ by Z .

If $X^n \underset{\theta^{n-1}}{\overset{\theta^0}{\vdots}} X^{n-1}$ is a family of n parallel morphisms in $\underline{\underline{T}}$, also written $(\theta^0, \ldots, \theta^{n-1}) : X^n \Rrightarrow X^{n-1}$, its <u>simplicial kernel</u> $\ker((\theta^0, \ldots, \theta^{n-1}) : X^n \Rrightarrow X^{n-1})$ is defined to be the largest subalgebra K of ${}^{n+1}(X^n)$ for which

$K \underset{\pi^n}{\overset{\pi^0}{\vdots}} X^n \underset{\theta^{n-1}}{\overset{\theta^0}{\vdots}} X^{n-1}$ is the image of a functor from $\cdot \underset{\varepsilon_{n+1}^n}{\overset{\varepsilon_{n+1}^0}{\vdots}} \cdot \underset{\varepsilon_n^{n-1}}{\overset{\varepsilon_n^0}{\vdots}} \cdot$

to $\underline{\underline{T}}$. Thus $K = \{ (x_0,\ldots,x_n) \in {}^{n+1}(X^n) \mid x_0\theta^0 = x_1\theta^0 , x_1\theta^1 = x_2\theta^1 , \ldots, x_n\theta^0 = x_0\theta^{n-1}$, etc. $\}$. In particular $\ker((\theta^0) : X^1 \Longrightarrow X^0)$ is just the usual $\ker(\theta^0 : X^1 \to X^0)$. If

$$B^n \;\begin{array}{c} \xrightarrow{\varepsilon^0} \\ \cdots \\ \xrightarrow{\varepsilon^{n-1}} \\ \xleftarrow{\delta^0} \\ \cdots \\ \xleftarrow{\delta^{n-2}} \end{array}\; B^{n-1}$$

is the image of a functor from

$$\cdot \;\begin{array}{c} \xrightarrow{\varepsilon^0_n} \\ \cdots \\ \xrightarrow{\varepsilon^{n-1}_n} \\ \xleftarrow{\delta^0_{n-1}} \\ \cdots \\ \xleftarrow{\delta^{n-2}_{n-1}} \end{array}\; \cdot$$

to $\underline{\underline{T}}$, and $K = \ker((\varepsilon^0,\ldots,\varepsilon^{n-1}) : B^n \Longrightarrow B^{n-1})$, defining $0^j : B^n \to K$ for $j = 0,\ldots,n-1$ by $0^j : x \mapsto (x_0,\ldots,x_n)$ where

$$x_{n-i} = \begin{cases} x\varepsilon^i\delta^{j-1} , & i < j \\ x , & i = j , j+1 \\ x\varepsilon^{i-1}\delta^j , & i > j+1 \end{cases}$$

makes

$$K \;\begin{array}{c} \xrightarrow{\pi^0} \\ \cdots \\ \xrightarrow{\pi^n} \\ \xleftarrow{0^0} \\ \cdots \\ \xleftarrow{0^{n-1}} \end{array}\; B^n \;\begin{array}{c} \xrightarrow{\varepsilon^0} \\ \cdots \\ \xrightarrow{\varepsilon^{n-1}} \\ \xleftarrow{\delta^0} \\ \cdots \\ \xleftarrow{\delta^{n-2}} \end{array}\; B^{n-1}$$

the image of a functor from

$$\cdot \;\begin{array}{c} \xrightarrow{\varepsilon^0_{n+1}} \\ \cdots \\ \xrightarrow{\varepsilon^n_{n+1}} \\ \xleftarrow{\delta^0_n} \\ \cdots \\ \xleftarrow{\delta^{n-1}_n} \end{array}\; \cdot \;\begin{array}{c} \xrightarrow{\varepsilon^0_n} \\ \xrightarrow{\varepsilon^{n-1}_n} \\ \xleftarrow{\delta^0_{n-1}} \\ \xleftarrow{\delta^{n-2}_{n-1}} \end{array}\; \cdot$$

to $\underline{\underline{T}}$.

In particular $0^0 : X^1 \to \ker(\theta^0 : X^1 \to X^0)$ is just the usual diagonal embedding.

Removing all operations involving lists of more than n symbols from Δ leaves _the simplicial category truncated at_ n . A functor from a truncated simplicial category to $\underline{\underline{T}}$ is called a _truncated simplicial object_. Any truncated simplicial object may be extended to a simplicial object by tacking on simplicial kernels with

π's and 0's repeatedly. Morphisms of truncated simplicial objects (i.e. natural transformations of the corresponding functors) may similarly be extended to full simplicial maps. One thus abuses language by forgetting the word "truncated" and imagining appropriate extensions. As an extreme case of this one could regard a $\underline{\underline{T}}$-algebra R as the simplicial object

$$- - - \quad {}^2R \;\underset{0^0}{\overset{\pi^0,\ \pi^1}{\rightrightarrows}}\; R \xrightarrow{\ {}^1R\ } R \quad .$$

A simplicial object B^* is said to be <u>surjacent on</u> B^n if $(b_0,\ldots,b_n) \in \ker((\varepsilon_n^0,\ldots,\varepsilon_n^{n-1}) : B^n \Longrightarrow B^{n-1}) \Rightarrow \exists\, b^{n+1} \in B^{n+1}$. $\forall\ 0 \leqslant i \leqslant n$, $b^{n+1}\varepsilon^i_{n+1} = b^i$. Thus (the extension of) a simplicial object B^* truncated at B^n is surjacent on B^i for all $i \geqslant n$. Also, B^* is surjacent on B^0 iff $\varepsilon : B^1 \to B^0$ surjects.

Now for a concept which is fundamental to this chapter :

611 DEFINITION A simplicial object $B^* = B^2 \underset{d}{\overset{e^0,\ e^1}{\rightrightarrows}} B^1 \xrightarrow{\ e\ } B^0$ is said to be <u>seeded</u> if

(0) e surjects

(1) B^* is surjacent on B^1

(2) $\ker e^0 = \eta(\ker e^1)$.

If X is a $\underline{\underline{T}}$-algebra, congruences E^i on $\underline{\underline{T}}(X,B^2)$ are defined by $f\, E^i\, g \Leftrightarrow fe^i = ge^i$, for $i = 0,1$. Then E^0 centralises E^1 .

Let $B^2 \underset{d}{\overset{e^0}{\underset{e^1}{\rightrightarrows}}} B^1 \xrightarrow{e} B^0$ be a seeded simplicial object. Let $C = \{ c \in B^2 \mid ce^0 = ce^1 \}$. Define V on C by $c\, V\, c' \Leftrightarrow (ce^0d,c)\,(\ker e^0 \circ \ker e^1 | \ker e^0 \cap \ker e^1)\,(c'e^0d,c')$. Then $C^V \to B^0 ;\ c^V \mapsto ce^0e$ is a module over B^0 , and so over B^* . It is called the module <u>grown</u> by B^* . There is an isomorphism of B^0-modules $C^V \to (\ker e^0 \cap \ker e^1)^{(\ker e^0 \circ \ker e^1 | \ker e^0 \cap \ker e^1)}$; $c^V \mapsto (ce^0d,c)^{(\ker e^0 \circ \ker e^1 | \ker e^0 \cap \ker e^1)}$. Note that

$c\ (V \cap \ker e^1)\ c' \quad \Rightarrow$

$(ce^0d,c)\ (\ker e^0 \cap \ker e^1 | \ker e^0 \cap \ker e^1)\ (c'e^0d,c') \quad \Rightarrow$

$(c,c')\ (\ker e^0 \cap \ker e^1 | \ker e^0 \cap \ker e^1)\ (ce^0d,c'e^0d)$, and $ce^0d = ce^0de^0d = c'e^0de^0d = c'e^0d$, so $c = c'$.

Let α be a congruence on a $\underline{\underline{T}}$-algebra T . Define $e : T^{\eta(\alpha)} \to T^{\alpha \circ \eta(\alpha)} ;\ t^\alpha \mapsto t^{\alpha \circ \eta(\alpha)}$, $d : T^{\eta(\alpha)} \to \alpha^{(\eta(\alpha)|\alpha)}$;

$t^{\eta(\alpha)} \mapsto (t,t)^{(\eta(\alpha)|\alpha)}$, and $e^i : \alpha^{(\eta(\alpha)|\alpha)} \to T^{\eta(\alpha)}$; $(t_0,t_1)^{(\eta(\alpha)|\alpha)} \mapsto t_i^{\eta(\alpha)}$ for $i = 0,1$. These are clearly well-defined. By Proposition 229(iii), $\ker e^0 = \eta(\ker e^1)$. Then

$$\alpha^{(\eta(\alpha)|\alpha)} \underset{\xleftarrow{\ d\ }}{\overset{e^0,\ e^1}{\rightrightarrows}} T^{\eta(\alpha)} \xrightarrow{\ e\ } T^{\alpha \circ \eta(\alpha)}$$

is a seeded simplicial object said to be planted by α on T . It grows the module $(\alpha \cap \eta(\alpha))^{(\alpha \circ \eta(\alpha)|\alpha \cap \eta(\alpha))}$.

For a $\underline{\underline{T}}$-algebra R , the simplicial object

$$- - - \quad {}^2R \underset{\xleftarrow{\ 0^0\ }}{\overset{\pi^0,\ \pi^1}{\rightrightarrows}} R \xrightarrow{\ {}^1R\ } R$$

has already been mentioned. There is a less trivial way of associating a simplicial object with R which plays a major role in what follows. Define RG to be the free $\underline{\underline{T}}$-algebra on the set $\{ \{r\} \mid r \in R \}$. Define $\underline{\underline{T}}$-morphisms $\varepsilon_n^j : RG^n \to RG^{n-1}$;

$$\{\{..\{\{\{..\{\{r_{i_1..i_{n-1}}\}\omega_{i_2..i_{n-1}}\}..\}\omega_{i_{j+1}..i_{n-1}}\}\omega_{i_{j+2}..i_{n-1}}\}..\}\omega\} \mapsto$$
$$\{\{..\{\{..\{\{r_{i_1..i_{n-1}}\}\omega_{i_2..i_{n-1}}\}..\}\omega_{i_{j+1}..i_{n-1}}\omega_{i_{j+2}..i_{n-1}}\}..\}\omega\} .$$

In particular ε_n^0 deletes the innermost brackets and ε_n^{n-1} the outermost. Define $\underline{\underline{T}}$-morphisms $\delta_n^j : RG^n \to RG^{n+1}$;

$$\{\{..\{\{\{..\{\{r_{i_1..i_{n-1}}\}\omega_{i_2..i_{n-1}}\}..\}\omega_{i_{j+1}..i_{n-1}}\}\omega_{i_{j+2}..i_{n-1}}\}..\}\omega\} \mapsto$$

$\{\{..\{\{\{\{..\{\{r_{i_1..i_{n-1}}\}\omega_{i_2..i_{n-1}}\}..\}\omega_{i_{j+1}..i_{n-1}}\}\}\omega_{i_{j+2}..i_{n-1}}\}..\}\omega\}$.

In particular δ_n^0 repeats the innermost brackets and δ_n^{n-1} the outermost. This defines a simplicial object RG^* for any R in $\underline{\underline{T}}$. There are also mappings of sets (although not $\underline{\underline{T}}$-morphisms) $h_n : RG^n \to RG^{n+1}$ for all natural numbers n defined by $xh_n = \{x\}$ for all x in RG^n . These satisfy identities such as $h_n\varepsilon_{n+1}^n = 1$, $h_n\varepsilon_{n+1}^0 = \varepsilon_n^0 h_{n-1}$.

612 PROPOSITION RG^* is surjacent on R , RG , RG^2 , RG^3 .

Proof. (a) Surjacency on R :

$\forall\, r \in R$, $rh_0\varepsilon = \{r\}\varepsilon = r$.

(b) Surjacency on RG :

Suppose $(x^0,x^1) \in \ker(\ \varepsilon : RG \to R\ )$. Let $x = (x^0\delta, x^0h_1, x^1h_1)P \in RG^2$. Then $x\varepsilon^0 = (x^0, x^0\varepsilon h_0, x^1\varepsilon h_0)P = x^0$ and $x\varepsilon^1 = (x^0,x^0,x^1)P = x^1$.

(c) Surjacency on RG^2 :

Suppose $(x^0,x^1,x^2) \in \ker((\varepsilon^0,\varepsilon^1) : RG^2 \rightrightarrows RG\ )$. There is the following "multiplication table" denoting the actions of the maps at the tops of the columns on the elements at the left of the rows :

	ε^0	ε^1	ε^2
$y^0 = x^0\delta^0$	x^0	x^0	$x^0\varepsilon^1\delta$
$y^1 = (x^0\delta^1, x^0h_2, x^0\varepsilon^1\delta h_2)P$	$(x^0\varepsilon^0\delta, x^0\varepsilon^0h_1, x^0\varepsilon^1h_1)P$	x^0	$x^0\varepsilon^1\delta$
$y^2 = (x^1\delta^1, x^1h_2, x^2h_2)P$	$(x^0\varepsilon^0\delta, x^0\varepsilon^0h_1, x^0\varepsilon^1h_1)P$	x^1	x^2

Thus if $y = (y^0,y^1,y^2)P \in RG^3$, $y\varepsilon^i = x^i$ for $i = 0,1,2$.

FIGURE 613

	ε^0	ε^1	ε^2	ε^3
$y^0 = x^0\delta^0$	x^0	x^0	$x^0\varepsilon^1\delta^0$	$a^0 = x^0\varepsilon^2\delta^0$
$y^1 = (x^0\delta^1, x^0\delta^2, x^0\varepsilon^1\delta^0\delta^2)P$	$(x^0\varepsilon^0\delta^0, x^0\varepsilon^0\delta^1, x^0\varepsilon^1\delta^1)P$	x^0	$x^0\varepsilon^1\delta^0$	$a^1 = (x^0\varepsilon^2\delta^1, x^0, x^0\varepsilon^1\delta^0)P$
$y^2 = (x^1\delta^1, x^1\delta^2, x^2\delta^2)P$	$(x^0\varepsilon^0\delta^0, x^0\varepsilon^0\delta^1, x^0\varepsilon^1\delta^1)P$	x^1	x^2	$a^2 = (x^1\varepsilon^2\delta^1, x^1, x^2)P$
$z^0 = x^0\varepsilon^2\delta^0h_3$	$x^0\varepsilon^2h_2$	$x^0\varepsilon^2h_2$	$x^0\varepsilon^2\varepsilon^1\delta h_2$	a^0
$z^1 = (x^0\varepsilon^2\delta^1h_3, x^0h_3, x^0\varepsilon^1\delta^0h_3)P$	$(x^0\varepsilon^2\varepsilon^0\delta h_2, x^0\varepsilon^0h_2, x^0\varepsilon^1h_2)P$	$x^0\varepsilon^2h_2$	$x^0\varepsilon^2\varepsilon^1\delta h_2$	a^1
$z^2 = (x^1\varepsilon^2\delta^1h_3, x^1h_3, x^2h_3)P$	$(x^0\varepsilon^2\varepsilon^0\delta h_2, x^0\varepsilon^0h_2, x^0\varepsilon^1h_2)P$	$x^1\varepsilon^2h_2$	$x^2\varepsilon^2h_2$	a^2
$t^0 = (y^0, y^1, y^2)P$	x^0	x^1	x^2	$(a^0, a^1, a^2)P$
$t^1 = (z^0, z^1, z^2)P$	$x^0\varepsilon^2h_2$	$x^1\varepsilon^2h_2$	$x^2\varepsilon^2h_2$	$(a^0, a^1, a^2)P$
$t^2 = x^3h_3$	$x^0\varepsilon^2h_2$	$x^1\varepsilon^2h_2$	$x^2\varepsilon^2h_2$	x^3

(d) Surjacency on RG^3 :

Suppose $(x^0,x^1,x^2,x^3) \in \ker((\varepsilon^0,\varepsilon^1,\varepsilon^2): RG^3 \Rrightarrow RG^2)$. There is the multiplication table Figure 613. Then if $t = (t^0,t^1,t^2)P \in RG^4$, $t\varepsilon^i = x^i$ for $i = 0,1,2,3$.]

In fact, RG* is surjacent on RG^n for all natural numbers n , but only the first four cases above will be needed.

The main purpose of RG* is to define cohomology. Let $Z \to R$ be an R-module, and regard it as a module over RG* . For each natural number n , $Der(RG^n,Z)$ is an abelian group. Define homomorphisms $d_n : Der(RG^n,Z) \to Der(RG^{n+1},Z)$; $f \mapsto \sum_{i=0}^{n}(-)^i\varepsilon^i_{n+1}f$. The suffix n is usually dropped from d_n . These d are known as <u>coboundary homomorphisms</u>. $IM(d_n) = \{ f \in Der(RG^{n+1},Z) \mid \exists\, g \in Der(RG^n,Z).\ gd_n = f \}$ is the set of <u>coboundaries</u> in $Der(RG^{n+1},Z)$, and $KER(d_n) = \{ f \in Der(RG^n,Z) \mid fd_n = 0 \}$ is the set of <u>cocycles</u> in $Der(RG^n,Z)$. Since $d_{n-1}d_n = 0$ for positive integers n , $IM(d_{n-1}) \leqslant KER(d_n)$. The cosets of $IM(d_{n-1})$ in $KER(d_n)$ are known as <u>cohomology classes</u>, and the quotient group $KER(d_n)/IM(d_{n-1})$ is called the $(n-1)$-th <u>cohomology group</u> $H^{n-1}(R,Z)$ <u>of</u> R <u>with coefficients in</u> Z .

For all natural numbers n , define $Der_0(RG^n,Z) =$

$\{ f \in Der(RG^n, Z) \mid \forall\ 0 \leqslant i < n-1\ ,\ \delta^i f = 0 \in Der(RG^{n-1}, Z) \}$.

Derivations in $Der_0(RG^n, Z)$ are said to be <u>normalised</u>.

614 PROPOSITION The cohomology classes in $H^1(R,Z)$ and $H^2(R,Z)$ may be represented by normalised cocycles.

<u>Proof.</u> (a) Classes in $H^1(R,Z)$:

Let $f \in KER(d_2)$, so that $\varepsilon^0 f - \varepsilon^1 f + \varepsilon^2 f = 0 \in Der(RG^3, Z)$, and let $f' = f - \varepsilon^0 \delta f \in Der_0(RG^2, Z)$. Then $(\delta f)d_1 = \varepsilon^0 \delta f - \varepsilon^1 \delta f =$ $\varepsilon^0 \delta f - \varepsilon^1 \delta f + \delta^0 \varepsilon^0 f - \delta^0 \varepsilon^1 f + \delta^0 \varepsilon^2 f = \varepsilon^0 \delta f = f - f'$, so that f and f' are in the same cohomology class.

(b) Classes in $H^2(R,Z)$:

Let $f \in KER(d_3)$, so that $\varepsilon^0 f - \varepsilon^1 f + \varepsilon^2 f - \varepsilon^3 f = 0 \in Der(RG^4, Z)$, and let $f' = f - \varepsilon^0 \delta^0 f + \varepsilon^0 \delta^1 f - \varepsilon^1 \delta^1 f \in Der_0(RG^3, Z)$. Then

$$\begin{aligned}(\delta^0 f)d_2 &= \varepsilon^0\delta^0 f - \varepsilon^1\delta^0 f + \varepsilon^2\delta^0 f \\ &= \varepsilon^0\delta^0 f - \varepsilon^1\delta^0 f + \varepsilon^2\delta^0 f \\ &\quad + \delta^0\varepsilon^0 f - \delta^0\varepsilon^1 f + \delta^0\varepsilon^2 f - \delta^0\varepsilon^3 f \\ &\quad - \delta^2\varepsilon^0 f + \delta^2\varepsilon^1 f - \delta^2\varepsilon^2 f + \delta^2\varepsilon^3 f \\ &= \varepsilon^0\delta^0 f - \varepsilon^0\delta^1 f + \varepsilon^1\delta^1 f = f - f'\end{aligned}$$

, so that f and f' are in the same cohomology class.]

6.2 Obstructions to extensions.

This section considers the following problem :

Given a seeded simplicial object B^* in $\underline{\underline{T}}$ and an epimorphism $p^0 : R \to B^0$, when is there a $\underline{\underline{T}}$-algebra T with a congruence α on it planting B^* and with $p^0 : R = T^{\alpha} \to T^{\alpha \cdot \eta(\alpha)} = B^0$?

The solution is given by Theorem 623 in terms of the cohomology of R with coefficients in the module grown by B^* .

Associated with the $\underline{\underline{T}}$-algebra R is the simplicial object RG^* in $\underline{\underline{T}}$. The epimorphism $p^0 : R \to B^0$ then gives rise to a morphism $p^* : RG^* \to B^*$ of simplicial objects by the following diagram chase. Since RG is free in $\underline{\underline{T}}$ and $e : B^1 \to B^0$ surjects, there is a (not necessarily unique) $p^1 : RG \to B^1$ such that $p^1 e = \varepsilon p$. Define $u : B^2 \to \ker e \,;\, b \mapsto (be^0, be^1)$. Since B^* is surjacent on B , u surjects. Define $\underline{p}^2 : RG^2 \to \ker e \,;\, x \mapsto (x\varepsilon^0 p^1, x\varepsilon^1 p^1)$. Since u surjects and RG^2 is free, there is a (not necessarily unique) $p^2 : RG^2 \to B^2$ such that $p^2 u = \underline{p}^2$. Then $p^2 e^i = p^2 u \pi^i = \underline{p}^2 \pi^i = \varepsilon^i p^1$ for $i = 0,1$. Finally, since $B^3 = \ker((e^0, e^1) : B^2 \rightrightarrows B^1)$, there is a $p^3 : RG^3 \to B^3$ such that $p^3 e^i = \varepsilon^i p^2$ for $i = 0,1,2$:

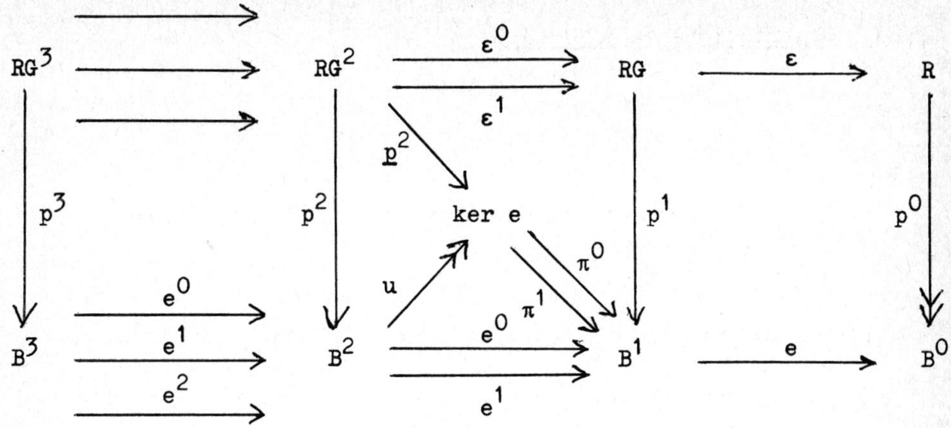

The module M grown by B^* becomes a module over RG^* . Since $(e^0e^0,e^1e^0,e^2e^0)P = e^2e^0 = e^0e^1 = (e^0e^1,e^1e^1,e^2e^1)P : B^3 \to B^1$, $B\ (e^0,e^1,e^2)P \subseteq C$. Let $D = (e^0,e^1,e^2)P^{\vee} : B^3 \to M$. It is a derivation in $\mathrm{Der}(B^3,M)$.

621 PROPOSITION $p^3D : RG^3 \to M$ is a cocycle in $\mathrm{Der}(RG^3,M)$.

<u>Proof.</u> Certainly, since $D : B^3 \to M$ is a derivation, $p^3D : RG^3 \to M$ is. It must be shown that $\sum_{i=0}^{3} (-)^i \varepsilon^i p^3 D = 0$ in $\mathrm{Der}(RG^4,M)$. Now
$((\varepsilon^0\varepsilon^0p^2,\varepsilon^0\varepsilon^1p^2,\varepsilon^0\varepsilon^2p^2)P,\varepsilon^0\varepsilon^0p^2)\ (E^0|E^1)\ (\varepsilon^0\varepsilon^2p^2,\varepsilon^0\varepsilon^1p^2)$,
$(\varepsilon^0\varepsilon^0p^2,(\varepsilon^1\varepsilon^0p^2,\varepsilon^1\varepsilon^1p^2,\varepsilon^1\varepsilon^2p^2)P)\ (E^0|E^1)\ (\varepsilon^1\varepsilon^1p^2,\varepsilon^1\varepsilon^2p^2)$, and
$(\varepsilon^0\varepsilon^1p^2,(\varepsilon^0\varepsilon^1p^2,\varepsilon^1\varepsilon^1p^2,\varepsilon^1\varepsilon^2p^2)P)\ (E^0|E^1)\ (\varepsilon^1\varepsilon^1p^2,\varepsilon^1\varepsilon^2p^2)$. Thus

$$(\varepsilon^0(\varepsilon^0p^2,\varepsilon^1p^2,\varepsilon^2p^2)P,\varepsilon^1(\varepsilon^0p^2,\varepsilon^1p^2,\varepsilon^2p^2)P)\ (E^0|E^1)\ (\varepsilon^0\varepsilon^2p^2,(\varepsilon^0\varepsilon^1p^2,\varepsilon^1\varepsilon^1p^2,\varepsilon^1\varepsilon^2p^2)P).$$

But both sides of this relation are in $E^0 \cap E^1$, and so it can be written as $(\varepsilon^0(\varepsilon^0p^2,\varepsilon^1p^2,\varepsilon^2p^2)P,\varepsilon^1(\varepsilon^0p^2,\varepsilon^1p^2,\varepsilon^2p^2)P)\ (E^0 \cdot E^1|E^0 \cap E^1)$

$(\varepsilon^0\varepsilon^2p^2,(\varepsilon^0\varepsilon^1p^2,\varepsilon^1\varepsilon^1p^2,\varepsilon^1\varepsilon^2p^2)P)$. Similarly

$(\varepsilon^3(\varepsilon^0p^2,\varepsilon^1p^2,\varepsilon^2p^2)P,\varepsilon^2(\varepsilon^0p^2,\varepsilon^1p^2,\varepsilon^2p^2)P)\ (E^0\circ E^1|E^0\cap E^1)$

$(\varepsilon^0\varepsilon^2p^2,(\varepsilon^0\varepsilon^1p^2,\varepsilon^1\varepsilon^1p^2,\varepsilon^1\varepsilon^2p^2)P)$. Thus $\varepsilon^3p^3D = (\varepsilon^0p^3D,\varepsilon^1p^3D,\varepsilon^2p^3D)P$ as required.]

Now p^3D has the form $p^3D = (p^3e^0,p^3e^1,p^3e^2)P^V = (\varepsilon^0p^2,\varepsilon^1p^2,\varepsilon^2p^2)P^V$. There was an arbitrariness in the choice of p^1 and p^2 above, and this may affect the value of the cocycle p^3D in $\mathrm{Der}(RG^3,M)$. However,

622 PROPOSITION The cohomology class determined by p^3D in $H^2(R,M)$ is independent of the arbitrariness in the choices for p^1 and p^2 .

<u>Proof.</u> Let p_i^1 , p_i^2 ($i = 0,1$) be alternative choices for p^1 , p . The proof consists of a diagram chase to construct a "simplicial homotopy" between the simplicial maps p_0^* and p_1^* . Define $\underline{h}^1 : RG \to \ker e\ ;\ x \mapsto (xp_0^1,xp_1^1)$. Since u surjects and RG is free, there is an $h^1 : RG \to B^2$ with $h^1u = \underline{h}^1$, and then $h^1e^i = h^1u\pi^i = \underline{h}^1\pi^i = p_i^1$, $i = 0,1$. Also $p_0^2e^0e = \varepsilon^0p_0^1e = \varepsilon^0p_1^1e = p_1^2e^0e = p_1^2e^1e$, so by a similar argument there is a map $v : RG^2 \to B^2$ with $ve^i = p_i^2e^i$, $i = 0,1$. Now consider the three maps p_0^2 , v , $\varepsilon^1h^1 : RG^2 \to B^2$. Since $p_0^2e^0 = ve^0$, $ve^1 = p_1^2e^1 = \varepsilon^1p_1^1 = \varepsilon^1h^1e^1$, $\varepsilon^1h^1e^0 = \varepsilon^1p_0^1 = p_0^2e^1$, and $B^3 = \ker((e^0,e^1): B^2 \rightrightarrows B^1\ )$, there is a map

$h_0^2 : RG^2 \to B^3$ with $h_0^2 e^0 = p_0^2$, $h_0^2 e^1 = v$, $h_0^2 e^2 = \varepsilon^1 h^1$. Similar consideration of $\varepsilon^0 h^1$, v , $p_1^2 : RG^2 \to B^2$ yields a map $h^2 : RG^2 \to B^3$ with $h_1^2 e^0 = \varepsilon^0 h^1$, $h_1^2 e^1 = v$, $h_1^2 e^2 = p_1^2$.

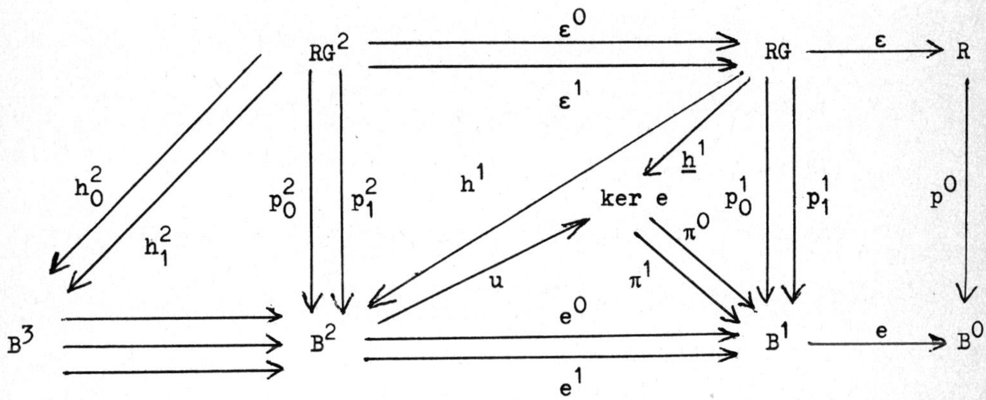

Now for $i = 0,1,2,$ $(\varepsilon^i p_0^2, \varepsilon^i v)\ (E^1|E^0)$

$(\varepsilon^i \varepsilon^1 h^1, (\varepsilon^i p_0^2, \varepsilon^i v, \varepsilon^i \varepsilon^1 h)P)$. Thus

$$((\varepsilon^0 p_0^2, \varepsilon^1 p_0^2, \varepsilon^2 p_0^2)P, (\varepsilon^0 v, \varepsilon^1 v, \varepsilon^2 v)P, (\varepsilon^0 \varepsilon^1 h^1, \varepsilon^1 \varepsilon^1 h^1, \varepsilon^2 \varepsilon^1 h^1)P)P =$$

$$((\varepsilon^0 p_0^2, \varepsilon^0 v, \varepsilon^0 \varepsilon^1 h^1)P, (\varepsilon^1 p_0^2, \varepsilon^1 v, \varepsilon^1 \varepsilon^1 h^1)P, (\varepsilon^2 p_0^2, \varepsilon^2 v, \varepsilon^2 \varepsilon^1 h^1)P)P .$$

Similarly $((\varepsilon^0 \varepsilon^0 h^1, \varepsilon^1 \varepsilon^0 h , \varepsilon^2 \varepsilon^0 h^1)P, (\varepsilon^0 v, \varepsilon^1 v, \varepsilon^2 v)P, (\varepsilon^0 p_1^2, \varepsilon^1 p_1^2, \varepsilon^2 p_1^2)P)P$

$$= ((\varepsilon^0 \varepsilon^0 h^1, \varepsilon^0 v, \varepsilon^0 p_1^2)P, (\varepsilon^1 \varepsilon^0 h^1, \varepsilon^1 v, \varepsilon^1 p_1^2)P, (\varepsilon^2 \varepsilon^0 h^1, \varepsilon^2 v, \varepsilon^2 p_1^2)P)P .$$

Hence $\sum_{i=0}^{2} (-)^i \varepsilon^i (h_0^2 D - h_1^2 D) = (\varepsilon^0 p_0^2, \varepsilon^1 p_0^2, \varepsilon^2 p_0^2)P^V - (\varepsilon^0 p_1^2, \varepsilon^1 p_1^2, \varepsilon^2 p_1^2)P^V$, showing as required that different possibilities for $p^3 D$ are all in the same cohomology class.]

This uniquely determined cohomology class of p^3D is called the <u>obstruction</u> of the morphism p^0. If the class of p^3D is the zero of $H^2(R,M)$, then p is said to be <u>unobstructed</u>. With this definition made, the solution to the problem of this section can be stated :

623 THEOREM Given a seeded simplicial object B^* in $\underline{\underline{T}}$ and an epimorphism $p^0 : R \to B^0$, there is a $\underline{\underline{T}}$-algebra T with a congruence α on it planting B^* and with $p^0 : R = T^{\alpha} \to T^{\alpha \circ \eta(\alpha)} = B^0$ if and only if p^0 is unobstructed.

<u>Proof.</u> Firstly, suppose there is a congruence α on a $\underline{\underline{T}}$-algebra T with $p^0 : R = T^{\alpha} \to T^{\alpha \circ \eta(\alpha)} = B^0$. α plants the bottom row of this commutative diagram

624

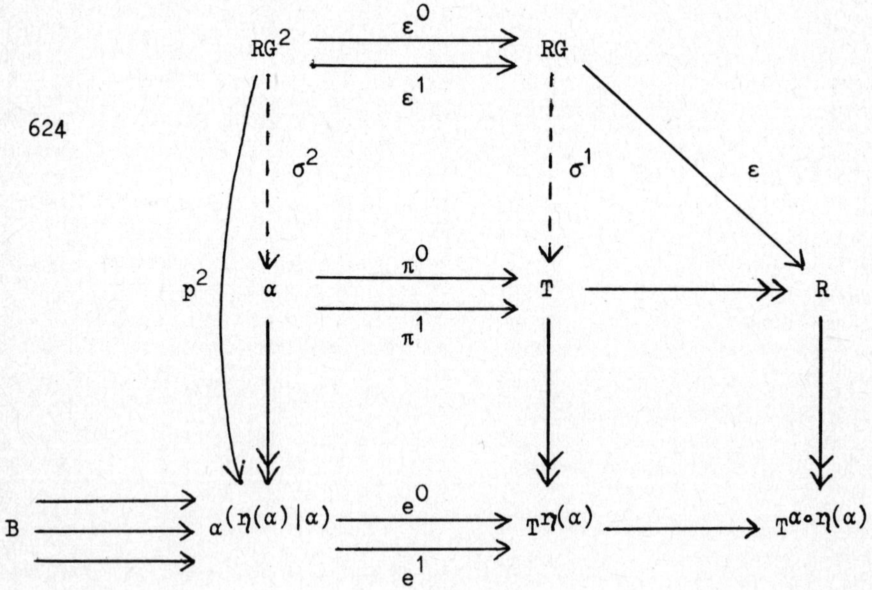

in which suitable σ^1 exists since RG is free, and σ^2 exists since $\alpha = \ker(T \to T^{\alpha})$. Call the morphisms down the respective columns of the diagram $p^2 = \sigma^2 \mathrm{nat}(\eta(\alpha)|\alpha)$, $p^1 = \sigma^1 \mathrm{nat}\eta(\alpha)$, and (as given) p^0 . Now $(\varepsilon^0\sigma^2,\varepsilon^1\sigma^2,\varepsilon^2\sigma^2)P\pi^0 = (\varepsilon^0\varepsilon^0,\varepsilon^1\varepsilon^0,\varepsilon^2\varepsilon^0)P\sigma^1 = \varepsilon^2\varepsilon^0\sigma^1 = \varepsilon^0\varepsilon^1\sigma^1 =$ $(\varepsilon^0\varepsilon^1,\varepsilon^1\varepsilon^1,\varepsilon^2\varepsilon^1)P\sigma^1 = (\varepsilon^0\sigma^2,\varepsilon^1\sigma^2,\varepsilon^2\sigma^2)P\pi^1$, so the obstruction of the morphism p^0 is $(\varepsilon^0 p^2,\varepsilon^1 p^2,\varepsilon^2 p^2)P^V = 0$ in $\mathrm{Der}(RG^3,(\alpha \cap \eta(\alpha))^{(\alpha\circ\eta(\alpha)|\alpha\cap\eta(\alpha))})$, as required.

Conversely, suppose given a seeded simplicial object B^* , an unobstructed epimorphism $p^0 : R \to B^0$, and hence choices for $p^i : RG^i \to B^i$, $i = 1,2,3$. Since p^0 is unobstructed, there is a derivation $\underline{c}$ in $\mathrm{Der}(RG^2,M)$ such that $(\varepsilon^0\underline{c},\varepsilon^1\underline{c},\varepsilon^2\underline{c})P =$ $(\varepsilon^0 p^2,\varepsilon^1 p^2,\varepsilon^2 p^2)P^V$. Since RG^2 is free, $\underline{c}$ lifts to $c : RG^2 \to C$. Define $\underline{p}^2 : RG^2 \to B^2$ by $(c,ce^0d)\ (E^0\circ E^1|E^0\cap E^1)\ (p^2,\underline{p}^2)$, with corresponding $\underline{p}^3 : RG^3 \to B^3$. $(p^0,p^1,\underline{p}^2,\underline{p}^3) : RG^* \to B^*$ is a simplicial map in $\underline{\underline{T}}$. Then $\underline{p}^3 D = (\varepsilon^0\underline{p}^2,\varepsilon^1\underline{p}^2,\varepsilon^2\underline{p}^2)P^V =$ $((\varepsilon^0 p^2,\varepsilon^1 p^2,\varepsilon^2 p^2)P^V,(\varepsilon^0 c,\varepsilon^1 c,\varepsilon^2 c)P^V,(\varepsilon^0 ce^0 d,\varepsilon^1 ce^0 d,\varepsilon^2 ce^0 d)P^V)P =$ $(\varepsilon^0 ce^0 d,\varepsilon^1 ce^0 d,\varepsilon^2 ce^0 d)P^V = 0$ in $\mathrm{Der}(RG^3,M)$. So without loss of generality $p^3 D$ may be assumed to be zero already. Under this assumption an appropriate algebra T will be constructed.

Let $Q = \{ (\{r_i\}\omega,b) \mid \{r_i\}\omega \in RG\ ,\ b \in B^2\ ,\ \{r_i\}\omega p^1 = be^1 \}$. Note that $(\{r_i\}\omega,b) \in Q \Rightarrow \{\{r_i\}\omega\}p^2 e^1 = \{\{r_i\}\omega\}\varepsilon^1 p^1 =$

$\{r_i\}\omega p^1 = be^1$, and that further $(\{r'_{i'}\}\omega',b') \in Q$, $r_i\omega = r'_{i'}\omega'$ $\Rightarrow$ $\{\{r_i\}\omega\}p^2e^0 = \{\{r_i\}\omega\}\varepsilon^0p^1 = \{r_i\omega\}p^1 = \{r'_{i'}\omega'\}p^1 = \{\{r'_{i'}\}\omega'\}p^2e^0$.

Let $W = \{ \ ((\{r_i\}\omega,b),(\{r'_{i'}\}\omega',b')) \in {}^2Q \mid r_i\omega = r'_{i'}\omega'$ and

$$(b,b')\ (\ker e^1|\ker e^0)\ (\{\{r_i\}\omega\}p^2,\{\{r'_{i'}\}\omega'\}p^2) \ \} .$$

Clearly $\hat{Q} \leqslant W$. Because $p^3D = 0$, W is a subalgebra of ${}^2Q$: let ω be an n-ary operation of $\underline{\underline{T}}$, and for $j = 1,\ldots,n$, let $((\{r_{ij}\}\omega_j,b_j),(\{r'_{i'j}\}\omega'_j,b'_j)) \in W$. Let $x = \{\{\{r_{ij}\}\omega_j\}\omega\}$, $x' = \{\{\{r'_{i'j}\}\omega'_j\}\omega\} \in RG^3$. Then $x\varepsilon^0 = \{\{r_{ij}\omega_j\}\omega\} = \{\{r'_{i'j}\omega'_j\}\omega\} = x'\varepsilon^0$. Now

$(x\varepsilon^1p^2,x\varepsilon^2p^2)\ (\ker e^0|\ker e^1)\ (x\varepsilon^0p^2,(x\varepsilon^0p^2,x\varepsilon^1p^2,x\varepsilon^2p^2)P)$,

$(x'\varepsilon^1p^2,x'\varepsilon^2p^2)\ (\ker e^0|\ker e^1)\ (x'\varepsilon^0p^2,(x'\varepsilon^0p^2,x'\varepsilon^1p^2,x'\varepsilon^2p^2)P)$, and since $(\varepsilon^0p^2,\varepsilon^1p^2,\varepsilon^2p^2)P^V = 0$ in $\mathrm{Der}(RG^3,M)$, $(x\varepsilon^0p^2,x\varepsilon^1p^2,x\varepsilon^2p^2)P$ V $(x'\varepsilon^0p^2,x'\varepsilon^1p^2,x'\varepsilon^2p^2)P$. But $V \cap \ker e^1 = \hat{C}$, so $(x\varepsilon^0p^2,x\varepsilon^1p^2,x\varepsilon^2p^2)P = (x'\varepsilon^0p^2,x'\varepsilon^1p^2,x'\varepsilon^2p^2)P$. Transitivity of $(\ker e^0|\ker e^1)$ then gives that $(x\varepsilon^1p^2,x\varepsilon^2p^2)\ (\ker e^0|\ker e^1)$ $(x'\varepsilon^1p^2,x'\varepsilon^2p^2)$, i.e. $(x\varepsilon^1p^2,x'\varepsilon^1p^2)\ (\ker e^1|\ker e^0)\ (x\varepsilon^2p^2,x'\varepsilon^2p^2)$. But $(b_j\omega,b'_j\omega)\ (\ker e^1|\ker e^0)\ (\{\{r_{ij}\}\omega_j\}\omega p^2,\{\{r'_{i'j}\}\omega'_j\}\omega p^2)$, so $(b_j\omega,b'_j\omega)\ (\ker e^1|\ker e^0)\ (\{\{r_{ij}\}\omega_j\omega\}p^2,\{\{r'_{i'j}\}\omega'_j\omega\}p^2)$, showing as required that $((\{r_{ij}\}\omega_j\omega,b_j\omega),(\{r'_{i'j}\}\omega'_j\omega,b'_j\omega))$ is in W . W becomes a congruence on Q .

Let $T = Q^W$. Define $\varphi : T \to R$; $(\{r_i\}\omega,b)^W \mapsto r_i\omega$ and let $\alpha = \ker\varphi$. Define $\theta : T \to B^1$; $(\{r_i\}\omega,b)^W \mapsto be^0$. Both these maps are well-defined, and there is the commutative square

$$
\begin{array}{ccc}
(\{r_i\}\omega,b)^W & \stackrel{\varphi}{\longmapsto} & r_i\omega \\
\big\downarrow \theta & & \big\downarrow p^0 \\
be^0 & \stackrel{e}{\longmapsto} & be^0e = be^1e = \{r_i\}\omega p^1e = \{r_i\}\omega\varepsilon p^0 = r_i\omega p^0
\end{array}
$$

$$
\begin{array}{ccc}
T & \stackrel{\varphi}{\longrightarrow} & R \\
\big\downarrow \theta & & \big\downarrow p^0 \\
B^1 & \stackrel{e}{\longrightarrow} & B^0
\end{array}
$$

Now θ surjects : let $b' \in B^1$, and consider $b'e$ in B^0 . Since p^0 surjects, there is an r in R such that $rp^0 = b'e$. Then $\{r\}p^1 \in B^1$ with $\{r\}p^1e = \{r\}\varepsilon p^0 = rp^0 = b'e$. Since B^* is seeded, there is a b in B^2 such that $be^0 = b'$, $be^1 = \{r\}p^1$. Then $(\{r\},b)$ is in Q and $\theta : (\{r\},b)^W \mapsto b'$.

The proof is completed by showing that $\ker \theta = \eta(\alpha)$. Suppose $(\{r_i\}\omega,b)^W \; \alpha \; (\{r'_{i'}\}\omega',b'')^W$. Then $r_i\omega = r'_{i'}\omega' \Rightarrow$ $\{\{r_i\}\omega\}p^2e^0 = \{\{r_i\}\omega\}\varepsilon^0p^1 = \{r_i\omega\}p^1 = \{r'_{i'}\omega'\}p^1 = \{\{r'_{i'}\}\omega'\}p^2e^0$. Let b' be specified by $(b',b'')\,(\ker e^1|\ker e^0)\,(\{\{r_i\}\omega\}p^2,\{r'_{i'}\}\omega'\}p^2)$. Then $(\{r_i\}\omega,b')\;W\;(\{r'_{i'}\}\omega',b'')$. The general pair in α can thus be written $((\{r_i\}\omega,b)^W,(\{r_i\}\omega,b')^W)$. Define $(\ker\theta|\alpha)$ by $((\{r_i\}\omega,b_1)^W,(\{r_i\}\omega,b_2)^W)\;(\ker\theta|\alpha)\;((\{r'_{i'}\}\omega',b'_1)^W,(\{r'_{i'}\}\omega',b'_2)^W) \Leftrightarrow$ $(b_1,b_2)\;(\ker e^0|\ker e^1)\;(b'_1,b'_2)$. By Corollary 224, $\ker\theta$

centralises α . If a congruence strictly containing $\ker e$ centralised α , then a congruence strictly containing $\ker e^0$ would centralise $\ker e^1$, contradictory to the seededness of B^* .]

6.3 Classifying extensions.

For a seeded simplicial object B^* growing a module M and an unobstructed epimorphism $p^0 : R \to B^0$ in $\underline{\underline{T}}$, Theorem 623 showed that there is a $\underline{\underline{T}}$-algebra T with a congruence α on it planting B^* and with $p^0 : R = T^\alpha \to T^{\alpha \circ \eta(\alpha)} = B^0$ the natural projection. This section classifies all such extensions $\alpha \rightrightarrows T \to R$. Two extensions, $\alpha_i \rightrightarrows T_i \to R$ for $i = 1,2$, are said to be <u>equivalent</u> if there is an isomorphism $T_1 \xrightarrow{\theta} T_2$ (both over R: $T_1 \searrow R \swarrow T_2$) of algebras over R .

Let $\underline{N}$ denote the class of equivalence classes planting B^* and inducing $p^0 : R \to B^0$. Let $\underline{S}$ denote the class of equivalence classes of extensions $\alpha \rightrightarrows T \to R$ with α self-centralising and an isomorphism $\alpha^{(\alpha|\alpha)} \longrightarrow M$ (both over R: $\alpha^{(\alpha|\alpha)} \searrow R \swarrow M$) of R-modules. Then $\underline{S}$ and $\underline{N}$ will be shown to be in 1-1 correspondence with the first cohomology group

$H^1(R,M)$.

Let $\alpha \rightrightarrows T \to R$ be a representative of an equivalence class in $\underline{S}$. One may construct a diagram like 624 for this, but with $\eta(\alpha)$ replaced by α . Then $p^2 \in \mathrm{Der}(RG^2,\alpha^{(\alpha|\alpha)})$. Because of the commuting of the diagram, $(\varepsilon^0 p^2,\varepsilon^1 p^2,\varepsilon^2 p^2)P = 0$ in $\mathrm{Der}(RG^3,\alpha^{(\alpha|\alpha)})$. Thus p^2 is a cocycle. Note also that p^2 is normalised : it lies in $\mathrm{Der}_0(RG^2,\alpha^{(\alpha|\alpha)})$. Let σ_i^1 , σ_i^2 be possible choices of the mappings σ^1 , σ^2 respectively in the diagram, for $i = 0,1$. Let p_i^2 for $i = 0,1$ be the corresponding choices of p^2 . As the images of an element of RG under σ_0^1 , σ_1^1 get mapped to the same element of R , and as α is the kernel of $T \to R$, there is a $\underline{\underline{T}}$-morphism $h : RG \to \alpha$ with $h\pi^i = \sigma_i^1$ for $i = 0,1$. Then $\varepsilon^i h = \sigma_i^2$ for $i = 0,1$, so $\varepsilon^0 h\ \mathrm{nat}(\alpha|\alpha) - \varepsilon^1 h\ \mathrm{nat}(\alpha|\alpha) = p_0^2 - p_1^2$. Thus the different choices for p^2 are in the same cohomology class in $H^1(R,\alpha^{(\alpha|\alpha)})$. This has constructed a function $\theta : \underline{S} \to H^1(R,\alpha^{(\alpha|\alpha)})$.

Conversely, suppose given an R-module $\pi_Z : Z \to R$ and an element of $H^1(R,Z)$ represented, using Proposition 614, by a normalised cocycle ρ . Let $\underline{A} = \{(x,z) \in RG \times Z \mid x\varepsilon = z\pi_Z\}$, i.e.

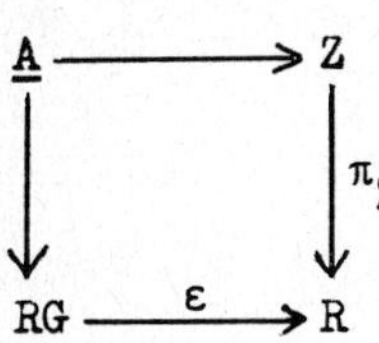

is a pullback. Define a relation U on $\underline{A}$ by

$(x_0,z_0)\ U\ (x_1,z_1)$ iff $\exists\ y \in RG^2$. $y\varepsilon^0 = x_0$, $y\varepsilon^1 = x_1$, and $z_1 - z_0 = y\rho$. Clearly $U \leqslant {}^2\underline{A}$. For any (x,z) in $\underline{A}$, $x\delta\varepsilon^0 = x$, $x\delta\varepsilon^1 = x$, and $z - z = 0 = x\delta\rho$ since ρ is normalised. Thus $\hat{\underline{A}} \leqslant U \leqslant {}^2\underline{A}$: U is a congruence on $\underline{A}$. Let $A = \underline{A}^U$. There is a well-defined morphism $A \to R$; $(x,z)^U \mapsto x\varepsilon$. Let α be its kernel.

Suppose there are x , x' in RG , y , y' in RG^2 such that

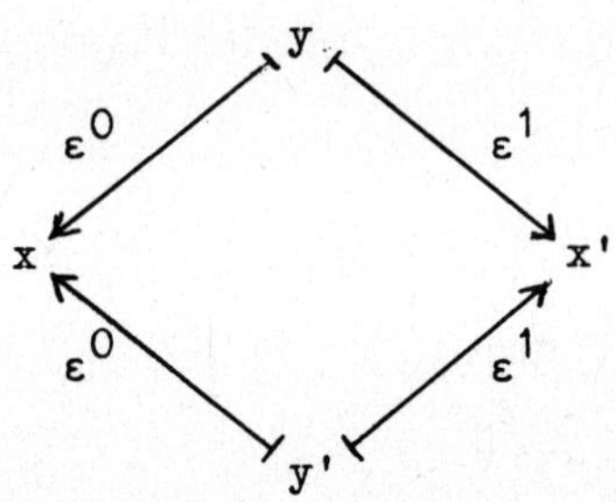

.

Consider the triple $(y,y',x'\delta)$. $y\varepsilon^0 = x = y'\varepsilon^0$, $y'\varepsilon^1 = x' = x'\delta\varepsilon^1$, and $x'\delta\varepsilon^0 = x' = y\varepsilon^1$, so $(y,y',x'\delta) \in \ker((\varepsilon^0,\varepsilon^1) : RG^2 \Rrightarrow RG)$. Thus there is an element w of RG^3 such that $w\varepsilon^0 = y$, $w\varepsilon^1 = y'$, and $w\varepsilon^2 = x'\delta$. Since ρ represents a cocycle, and is in $\mathrm{Der}_0(RG^2,Z)$, $y\rho - y'\rho = y\rho - y'\rho + x'\delta\rho = (w\varepsilon^0\rho, w\varepsilon^1\rho, w\varepsilon^2\rho)P = 0$. So given (x,z) in $\underline{A}$ and x' in $x^{\ker\varepsilon}$, there is at most one z' such that $(x,z)\ U\ (x',z')$. Conversely, if (x,x') is in $\ker\varepsilon$, there is a y in RG^2 such that $y\varepsilon^0 = x$, $y\varepsilon^1 = x'$. Then (x,z) U $(x',z+y\rho)$. Summarising,

631 $\forall\ (x,z) \in \underline{A}\ ,\ \forall\ x' \in x^{\ker\varepsilon}\ ,\ \exists_1\ z'\ .\ (x,z)\ U\ (x',z')\ .$

For $(x,z)^U\ \alpha\ (x',z")^U$, one can by 631 pick a unique z' such that $(x',z")\ U\ (x,z')$. Using this, define $(\alpha|\alpha)$ on α by $((x,z_{00})^U,(x,z_{10})^U)\ (\alpha|\alpha)\ ((x,z_{01})^U,(x,z_{11})^U) \Leftrightarrow z_{00} - z_{10} = z_{01} - z_{11}$. This is clearly a congruence on α with which it centralises itself, and $\alpha^{(\alpha|\alpha)} \to Z\ ;\ ((x,z)^U,(x,z')^U)^{(\alpha|\alpha)} \mapsto z - z'$ is an isomorphism of R-modules. Thus a function $\varphi : H^1(R,Z) \to \underline{S}$ has been constructed.

632 THEOREM The equivalence classes in $\underline{S}$ are in 1-1 correspondence with the elements of $H^1(R,M)$.

Proof. The above notation is used.

(a) $\varphi\theta$ is the identity mapping on $H^1(R,Z)$:

For any y in RG^2 , $(y\varepsilon^0,0)\ U\ (y\varepsilon^1,y\rho)$. Then with the notation of diagram 624, $yp^2 = ((y\varepsilon^0,0)^U,(y\varepsilon^1,0)^U)^{(\alpha|\alpha)} = ((y\varepsilon^1,y\rho)^U,(y\varepsilon^1,0)^U)^{(\alpha|\alpha)}$, which maps to $y\rho$ under the isomorphism $\alpha^{(\alpha|\alpha)} \to Z$ of R-modules.

(b) $\theta\varphi$ is the identity mapping on $\underline{S}$:

Let $\alpha \rightrightarrows T \xrightarrow{\pi} R$ be a representative of a class in $\underline{S}$. Let p^1 be as in diagram 624, and pick $p^2 : RG^2 \to \alpha^{(\alpha|\alpha)}$; $y \mapsto (y\varepsilon^0 p^1, y\varepsilon^1 p^1)^{(\alpha|\alpha)}$. Define a function $(RG \times_R \alpha^{(\alpha|\alpha)})^U \to T$; $(x,(t,t')^{(\alpha|\alpha)})^U \mapsto u$ where $(u,xp^1)\ (\alpha|\alpha)\ (t,t')$. This is well-defined, for if $(x,(t,t')^{(\alpha|\alpha)})\ U\ (x',(t,t'')^{(\alpha|\alpha)})$, say with $y\varepsilon^0 = x$, $y\varepsilon^1 = x'$, $y\rho = (xp^1, x'p^1)^{(\alpha|\alpha)}$, $(t,t'')^{(\alpha|\alpha)} = (t,t')^{(\alpha|\alpha)} + y\rho$, $(u,xp^1)\ (\alpha|\alpha)\ (t,t')$, $(u',x'p^1)\ (\alpha|\alpha)\ (t,t'')$, then

$$
\begin{aligned}
(u,u')^{(\alpha|\alpha)} &= (u,xp^1)^{(\alpha|\alpha)} + (xp^1,x'p^1)^{(\alpha|\alpha)} + (x'p^1,u')^{(\alpha|\alpha)} \\
&= (t,t')^{(\alpha|\alpha)} + y\rho + (t'',t)^{(\alpha|\alpha)} = (t,t)^{(\alpha|\alpha)} , \text{ so that}
\end{aligned}
$$

$u = u'$. The function has the two-sided inverse over R $T \to (RG \times_R \alpha^{(\alpha|\alpha)})^U$; $t \mapsto (\{t\pi\},(t,\{t\pi\}p^1)^{(\alpha|\alpha)})^U$. This function is a $\underline{\underline{T}}$-morphism, for consideration of $\{\{t_i\pi\}\omega\}$ in RG^2 shows that $(\{t_i\pi\omega\},(t_i\omega,\{t_i\pi\omega\}p^1)^{(\alpha|\alpha)})\ U\ (\{t_i\omega\},(t_i\omega,\{t_i\pi\}p^1\omega)^{(\alpha|\alpha)})$.]

The isomorphism $\underline{S} \cong H^1(R,M)$ of sets may be used to define an abelian group structure on $\underline{S}$, addition, + , being called Baer sum. The Baer sum of two extensions $\alpha_i \rightrightarrows T_i \to R$ in $\underline{S}$ for $i = 1,2$ with module isomorphism

$$
\begin{array}{ccc}
\alpha_1^{(\alpha_1|\alpha_1)} & \xrightarrow{\theta} & \alpha_2^{(\alpha_2|\alpha_2)} \\
& \searrow \quad \swarrow & \\
& R &
\end{array}
$$

may be realised more directly as the quotient of the pullback $T_1 \times_R T_2$ by $\{((t_1,t_2),(t_1',t_2')) \mid (t_i,t_i') \in \alpha_i \ , \ (t_1,t_1')^{(\alpha_1|\alpha_1)}\theta = (t_2',t_2)^{(\alpha_2|\alpha_2)} \}$.

The zero of $\underline{S}$ is the class of the split extension $M \to R$.

To produce a 1-1 correspondence between $\underline{S}$ and $\underline{N}$, operations $\underline{S} \times \underline{N} \to \underline{N}$; $(\underline{s},\underline{n}) \mapsto \underline{s} + \underline{n}$ and $\underline{N} \times \underline{N} \to \underline{S}$; $(\underline{n},\underline{n}') \mapsto \underline{n} - \underline{n}'$ will be defined such that $(\underline{n} - \underline{n}') + \underline{n}' = \underline{n}$ and $(\underline{s} + \underline{n}) - \underline{n} = \underline{s}$. Then for fixed $\underline{n}'$ in $\underline{N}$, the mappings $\underline{S} \to \underline{N}$; $\underline{s} \mapsto \underline{s} + \underline{n}'$ and $\underline{N} \to \underline{S}$; $\underline{n} \mapsto \underline{n} - \underline{n}'$ are inverses of each other.

Let $\beta \rightleftarrows S \to R$ be an element of the class $\underline{s}$ in $\underline{S}$, and let $\alpha \rightleftarrows T \to R$ represent $\underline{n}$ in $\underline{N}$. Let $(\alpha \cap \eta(\alpha))^{(\alpha|\alpha\cap\eta(\alpha))} \xrightarrow{\theta} \beta^{(\beta|\beta)}$ be the isomorphism of R-modules showing that $\underline{s}$ is in $\underline{S}$. Let T' be the quotient of the pullback $T \times_R S$ by the congruence $E = \{ ((t,s),(t',s')) \in {}^2(T \times_R S) \mid (t,t') \in \alpha \cap \eta(\alpha)$, $(s',s) \in \beta$, $(t,t')^{(\alpha|\alpha\cap\eta(\alpha))}\theta = (s',s)^{(\beta|\beta)} \}$. Note that

633 $\quad \forall\ (t,s)$, $(t',s') \in T \times_R S$ with $(t,t') \in \alpha$,

$$\exists_1\ t'' .\ (t'',s)\ E\ (t',s') .$$

t" may be defined as $t\ \alpha\ t' \Rightarrow s'\ \beta\ s$ by $(t'',t')^{(\alpha|\alpha\cap\eta(\alpha))}\theta = (s',s)^{(\beta|\beta)}$, and the uniqueness follows since $(t'',s)\ E\ (t^*,s) \Rightarrow (t'',t^*)^{(\alpha|\alpha\cap\eta(\alpha))}\theta = (s,s)^{(\beta|\beta)} \Rightarrow t'' = t^*$.

Define α' and η' on T' by $(t,s)^E\ \alpha'\ (t',s')^E$ iff $t\ \alpha\ t'$ and $(t,s)^E\ \eta'\ (t',s')^E$ iff $t\ \eta(\alpha)\ t'$. Define $(\eta'|\alpha')$ on α' (using 633) by $((t,s)^E,(t'',s)^E)\ (\eta'|\alpha')\ ((u,s^*)^E,(u'',s^*)^E) \Leftrightarrow (t,t'')\ (\eta(\alpha)|\alpha)\ (u,u'')$. Then η' is a congruence on T' centralising α' by $(\eta'|\alpha')$. If anything strictly larger than η' centralised α', something strictly larger than $\eta(\alpha)$ would centralise α, a contradiction. Thus $\eta' = \eta(\alpha')$. Note that for any (t,s), (t,s') in T' and (t,t'') in α, $((t,s)^E,(t'',s)^E)\ (\eta(\alpha')|\alpha')\ ((t,s')^E,(t'',s')^E)$. Now $j^2 : \alpha'^{(\eta(\alpha')|\alpha')} \to \alpha^{(\eta(\alpha)|\alpha)}$; $((t,s)^E,(t',s)^E)^{(\eta(\alpha')|\alpha')} \mapsto (t,t')^{(\eta(\alpha)|\alpha)}$, $j^1 : T'^{\eta(\alpha')} \to T^{\eta(\alpha)}$; $((t,s)^E)^{\eta(\alpha')} \mapsto t^{\eta(\alpha)}$, and $j^0 : T'^{\alpha'\circ\eta(\alpha')} \to T^{\alpha\circ\eta(\alpha)}$; $((t,s)^E)^{\alpha'\circ\eta(\alpha')} \mapsto t^{\alpha\circ\eta(\alpha)}$ form an isomorphism of the simplicial objects planted by $\alpha' \rightrightarrows T' \to R$ and $\alpha \rightrightarrows T \to R$ respectively. Since

$$
\begin{array}{ccc}
T^{\alpha} & \xrightarrow{\quad p^0 \quad} & T^{\alpha\circ\eta(\alpha)} \\
\| & & \uparrow \\
R & & j^0 \\
\| & & \\
T'^{\alpha'} & \longrightarrow & T'^{\alpha'\circ\eta(\alpha')}
\end{array},
$$

where the rows are the natural projections, commutes, it follows that the equivalence class of $\alpha' \rightrightarrows T' \to R$ is in $\underline{N}$. $\underline{s} + \underline{n}$ is defined to be this equivalence class.

Let $\alpha \rightrightarrows T \xrightarrow{\varphi} R$ be an element of the class $\underline{n}$ in $\underline{N}$ and let $\alpha' \rightrightarrows T' \xrightarrow{\varphi'} R$ be an element of the class $\underline{n}'$ in $\underline{N}$. Let the isomorphism of the simplicial objects planted by them be

$$\begin{array}{ccccc}
\alpha^{(\eta(\alpha)|\alpha)} & \rightrightarrows & T^{\eta(\alpha)} & \longrightarrow & T^{\alpha\circ\eta(\alpha)} \\
{\scriptstyle j^2}\downarrow & & {\scriptstyle j^1}\downarrow & & {\scriptstyle j^0}\downarrow \\
\alpha'^{(\eta(\alpha')|\alpha')} & \rightrightarrows & T'^{\eta(\alpha')} & \longrightarrow & T'^{\alpha'\circ\eta(\alpha')}
\end{array} \quad .$$

Let $j : T^{\alpha} \to T'^{\alpha'}$ be the corresponding isomorphism of the quotients. Let $W = \{ (t,t') \in T \times T' \mid t^{\eta(\alpha)}j^1 = t'^{\eta(\alpha')}$ and $t^{\alpha}j = t'^{\alpha'} \}$. Let S be the quotient of W by the congruence $E =$ $\{ ((t,t'),(u,u')) \mid (t,u) \in \alpha$, $(t',u') \in \alpha'$, $(t,u)^{(\eta(\alpha)|\alpha)}j^2 =$ $(t',u')^{(\eta(\alpha')|\alpha')} \}$. Define $\theta : S \to R$; $(t,t')^E \mapsto t\varphi = t'\varphi'$. This is clearly well-defined.

Given r in R , since φ and φ' surject, there are t in T , t' in T' such that $t\varphi = r = t'\varphi'$. Then $t^{\alpha\circ\eta(\alpha)}j^0 = t'^{\alpha'\circ\eta(\alpha')}$. Thus $\exists\ (u_0^{\eta(\alpha)},u_1^{\eta(\alpha)}) \in$ $(\alpha \circ \eta(\alpha))^2\mathrm{nat}\eta(\alpha)$. $u_0^{\eta(\alpha)} = t^{\eta(\alpha)}$, $u_1^{\eta(\alpha)}j^1 = t'^{\eta(\alpha')}$. Hence $\exists\ (v_0,v_1) \in \alpha$. $v_0^{\eta(\alpha)} = t^{\eta(\alpha)}$, $v_1^{\eta(\alpha)}j^1 = t'^{\eta(\alpha')}$. Define t'' by $(t,v_0)\ (\alpha|\eta(\alpha))\ (t'',v_1)$. Then $t\ \alpha\ t''$ implies $t''\varphi = t\varphi = r = t'\varphi'$. Also $t''^{\eta(\alpha)}j^1 = v_1^{\eta(\alpha)}j^1 = t'^{\eta(\alpha')}$. Thus $(t'',t') \in W$ and $(t'',t')^E\theta = r$. Hence θ surjects. Let $\beta = \ker(\ \theta : S \to R\ )$.

Suppose $(t,t')^E \ \beta \ (u,u')^E$, so that $u' \ \alpha' \ t'$.

Define $t"$ by $(u,t")^{(\eta(\alpha)|\alpha)}j^2 = (u',t')^{(\eta(\alpha')|\alpha')}$, which is possible since $u^{\eta(\alpha)}j^1 = u'^{\eta(\alpha')}$. Then $(u,u') \ E \ (t",t')$. Note that $t^{\eta(\alpha)}j^1 = t'^{\eta(\alpha')} = t"^{\eta(\alpha)}j^1$. Thus a general element $((t,t')^E,(u,u')^E)$ of β may be written as $((t,t')^E,(t",t')^E)$ with $t \ \alpha \cap \eta(\alpha) \ t"$. Then β centralises itself with $(\beta|\beta)$ defined by $((t,t')^E,(t",t')^E) \ (\beta|\beta) \ ((u,u')^E,(u",u')^E) \Leftrightarrow (t,t") \ (\alpha|\alpha \cap \eta(\alpha))$ $(u,u")$, and there is an isomorphism over R

$\beta^{(\beta|\beta)} \to (\alpha \cap \eta(\alpha))^{(\alpha|\alpha\cap\eta(\alpha))}$; $((t,t')^E,(t",t')^E)^{(\beta|\beta)} \mapsto$ $(t,t")^{(\alpha|\alpha\cap\eta(\alpha))}$. It follows that the equivalence class of $\beta \rightrightarrows S \to R$ is in $\underline{S}$. $\underline{n} - \underline{n}'$ is defined to be this equivalence class.

634 THEOREM $\underline{S}$ and $\underline{N}$ are in 1-1 correspondence.

<u>Proof.</u> Let $\alpha \rightrightarrows T \xrightarrow{\varphi} R$, $\alpha' \rightrightarrows T' \xrightarrow{\varphi'} R$ respectively be elements of the classes $\underline{n}$, $\underline{n}'$ in $\underline{N}$, with isomorphism j^* between the simplicial objects planted by them as before. Let $\beta \rightrightarrows S \xrightarrow{\psi} R$ be an element of the class $\underline{s}$ in $\underline{S}$, with $\theta : (\alpha \cap \eta(\alpha))^{(\alpha|\alpha\cap\eta(\alpha))} \to \beta^{(\beta|\beta)}$ **the isomorphism** of R-modules showing that $\underline{s}$ is in $\underline{S}$.

(a) $(\underline{n} - \underline{n}') + \underline{n}' = \underline{n}$:

The constructions given yield an algebra U representing $(\underline{n} - \underline{n}') + \underline{n}'$ which is a quotient of X =

$\{ (t,t',t") \in T \times T' \times T' \mid t^{\eta(\alpha)}j^1 = t'^{\eta(\alpha')} , t\varphi = t'\varphi' = t"\varphi' \}$.
Define a mapping $X \to T$; $(t,t',t") \mapsto t^*$ where t^* is defined by $(t,t^*)^{(\eta(\alpha)|\alpha)}j^2 = (t',t")^{(\eta(\alpha')|\alpha')}$ (possible since $t^{\eta(\alpha)}j^1 = t'^{\eta(\alpha')}$) . This factors through a mapping $\theta : U \to T$ which is an isomorphism of algebras over R .

(b) $(\underline{s} + \underline{n}) - \underline{n} = \underline{s}$:

The constructions given yield an algebra U representing $(\underline{s} + \underline{n}) - \underline{n}$ which is a quotient of $X = \{ (t,s,t') \in T \times S \times T \mid (t,t') \in \alpha \cap \eta(\alpha) , t\varphi = s\psi \}$. Define a mapping $X \to S$; $(t,s,t') \mapsto s'$ where s' is defined by $(t,t')^{(\alpha|\alpha\cap\eta(\alpha))}\theta = (s,s')^{(\beta|\beta)}$. This factors through a mapping $\theta : U \to S$ which is an isomorphism of algebras over R .]

It may be shown that for $\underline{s}'$ in $\underline{S}$, $(\underline{s} + \underline{s}') + \underline{n} = \underline{s} + (\underline{s}' + \underline{n})$, so that $\underline{S} \cong H^1(R,M)$ as an abelian group acts on $\underline{N}$ regularly and transitively.

6.4 Classifying obstructions.

Section 6.2 asked when a seeded simplicial object B^* and an epimorphism $p^0 : R \to B^0$ in $\underline{\underline{T}}$ yielded an algebra T having congruence α planting B^* with $p^0 : R = T^\alpha \to T^{\alpha\circ\eta(\alpha)} = B^0$. The answer was given by Theorem 623 in terms of the cohomology group $H^2(R,M)$, M being the module grown by B^* : such an extension T existed if and only if the obstruction of the morphism p^0 was the zero element of $H^2(R,M)$. The purpose of the current section is to show, under an appropriate hypothesis on $\underline{\underline{T}}$, that for any R-module Z every element of $H^2(R,Z)$ is the obstruction of some epimorphism. One can thus say that $H^2(R,Z)$ classifies obstructions : no smaller group contains all obstructions and no quotient group is fine enough to test whether an extension exists or not.

641 THEOREM Let R be in $\underline{\underline{T}}$ with the property that $\eta(\ker(\varepsilon : RG \to R)) = \widehat{RG}$. Let $\pi_Z : Z \to R$ be an R-module and let $\xi \in H^2(R,Z)$. Then there is a seeded simplicial object B^* and an epimorphism $p^0 : R \to B^0$ such that the module M grown by B^* is isomorphic to Z as an R-module and ξ is the obstruction of the morphism p^0.

<u>Proof.</u> By Proposition 614, ξ may be represented by a normalised

cocycle ρ in $Der_0(RG^3, Z)$. Let $\underline{A} = \{ (x,z) \in RG^2 \times Z \mid x\varepsilon^0\varepsilon = z\pi_Z \}$, i.e.

$$\begin{array}{ccc} \underline{A} & \longrightarrow & Z \\ \big\downarrow & & \big\downarrow \pi_Z \\ RG^2 & \xrightarrow{\varepsilon^0\varepsilon \,=\, \varepsilon^1\varepsilon} & R \end{array}$$

is a pullback. Define a relation U on $\underline{A}$ by $(x,z)\ U\ (x',z') \Leftrightarrow \exists\ y \in RG^3 .\ y\varepsilon^0 = x\ ,\ y\varepsilon^1 = x'\ ,\ y\varepsilon^2 = x\varepsilon^1\delta$, and $z' - z = y\rho$.

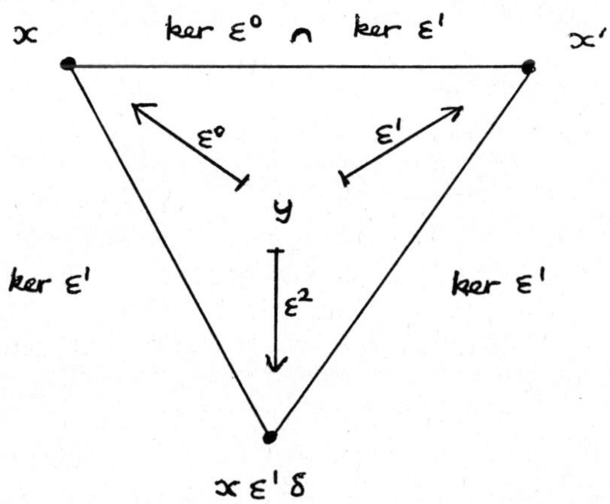

Since $\pi_Z : Z \to R$ is an R-module, $U \leqslant {}^2\underline{A}$. For any (x,z) in $\underline{A}$, $x\delta^0\varepsilon^0 = x\ ,\ x\delta^0\varepsilon^1 = x\ ,\ x\delta^0\varepsilon^2 = x\varepsilon^1\delta$, and $z - z = 0 = x\delta^0\rho$ since ρ is normalised. Thus $\hat{\underline{A}} \leqslant U$. U becomes a congruence on $\underline{A}$; let $A = \underline{A}^U$.

Suppose

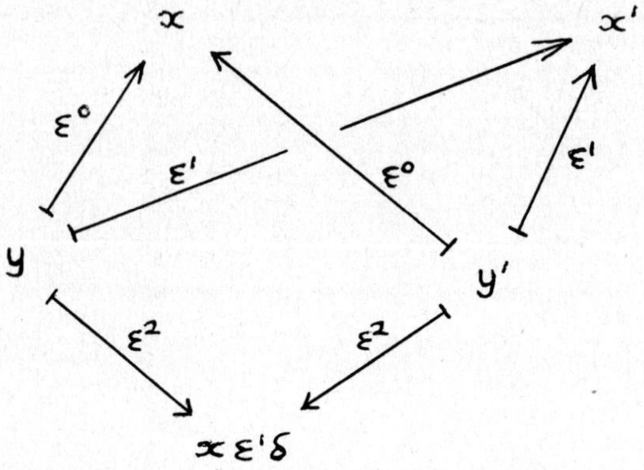

Then

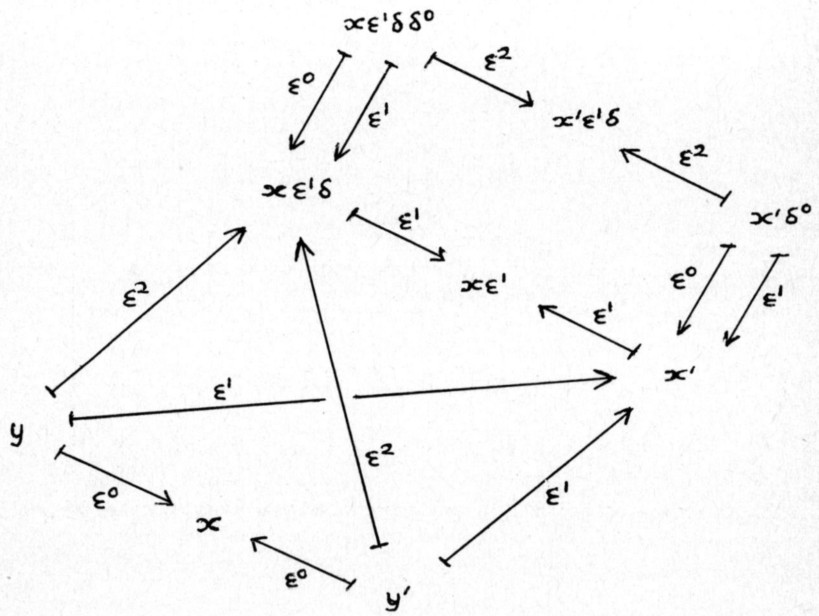

so that $(y, y', x'\delta^0, x\varepsilon^1\delta\delta^0) \in \ker((\varepsilon^0, \varepsilon^1, \varepsilon^2) : RG^3 \Rrightarrow RG^2)$.

Thus $\exists\, w \in RG^4$. $w\varepsilon^0 = y$, $w\varepsilon^1 = y'$, $w\varepsilon^2 = x'\delta^0$, $w\varepsilon^3 = x\varepsilon^1\delta\delta^0$. Since ρ is a normalised cocycle, it follows that $y\rho = y'\rho$. Thus, given $(x,z) \in \underline{A}$ and $x' \in x^{\ker\varepsilon^0 \cap \ker\varepsilon^1}$, there is at most one z' such that $(x,z)\ U\ (x',z')$. Conversely, suppose $(x,x') \in \ker\varepsilon^0 \cap \ker\varepsilon^1 \leqslant {}^2(RG^2)$. Since $x\varepsilon^0 = x'\varepsilon^0$, $x'\varepsilon^1 = x\varepsilon^1 = x\varepsilon^1\delta\varepsilon^1$, $x\varepsilon^1\delta\varepsilon^0 = x\varepsilon^1$, $\exists\, y \in RG^3$. $y\varepsilon^0 = x$, $y\varepsilon^1 = x'$, $y\varepsilon^2 = x\varepsilon^1\delta$. Then $(x,z)\ U\ (x',z+y\rho)$. Summarising,

642 $\quad \forall\, (x,z) \in \underline{A}$, $\forall\, x' \in x^{\ker\varepsilon^0 \cap \ker\varepsilon^1}$, $\exists_1\, z'$. $(x,z)\ U\ (x',z')$.

Let $E^i = \ker(\ A \to RG\ ;\ (x,z)^U \mapsto x\varepsilon^i\ )$, $i = 0,1$. Define $(E^1|E^0)$ on E^0 by $((x_{00},z_{00})^U,(x_{10},z_{10})^U)\ (E^1|E^0)\ ((x_{01},z_{01})^U,(x_{11},z_{11})^U) \Leftrightarrow \exists\, y_0\ , y_1 \in RG^3$. $z_{10} - z_{00} - y_0\rho = z_{11} - z_{01} - y_1\rho$, $y_0\varepsilon^2 = y_1\varepsilon^2$, and $\forall\, i\ , j \in \{0,1\}$, $y_i\varepsilon^j = x_{ji}$.

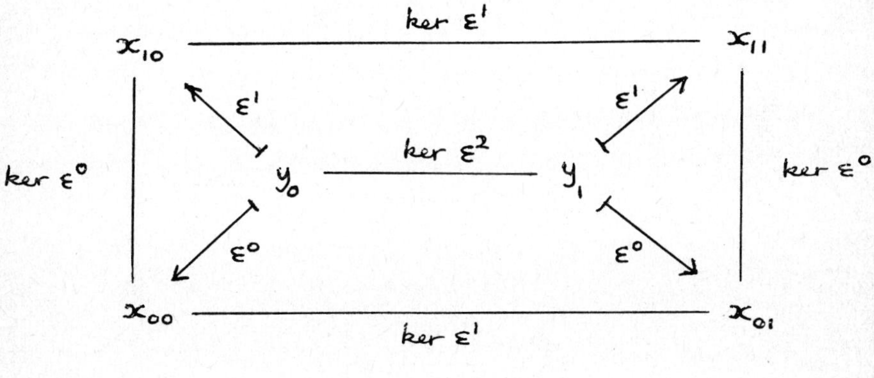

Since $\pi_Z : Z \to R$ is an R-module, $(E^1|E^0) \leqslant {}^2E^0$. Let $((x,z)^U,(x',z')^U) \in E^0$. Then $x\varepsilon^0 = x'\varepsilon^0$, $x'\varepsilon^1 = (x\varepsilon^1\delta\varepsilon^1, x\varepsilon^1, x'\varepsilon^1)P$

$= (x\varepsilon^1\delta, x, x')P\varepsilon^1$, and $(x\varepsilon^1\delta, x, x')P\varepsilon^0 = (x\varepsilon^1\delta\varepsilon^0, x\varepsilon^0, x'\varepsilon^0)P = x\varepsilon^1\delta\varepsilon^0$ $= x\varepsilon^1$, so that $(x, x', (x\varepsilon^1\delta, x, x')P) \in \ker((\varepsilon^0, \varepsilon^1) : RG^2 \Longrightarrow RG)$. Thus $\exists\, y \in RG^3$. $y\varepsilon^0 = x$, $y\varepsilon^1 = x'$. Hence $((x,z)^U, (x',z')^U)$ $(E^1|E^0)$ $((x,z)^U, (x',z')^U)$. So $\widehat{E^0} \leqslant (E^1|E^0) \leqslant {}^2E^0$: $(E^1|E^0)$ is a congruence on E^0 .

Suppose $((x,z)^U, (x,z)^U)$ $(E^1|E^0)$ $((x,z)^U, (x',z')^U)$, say

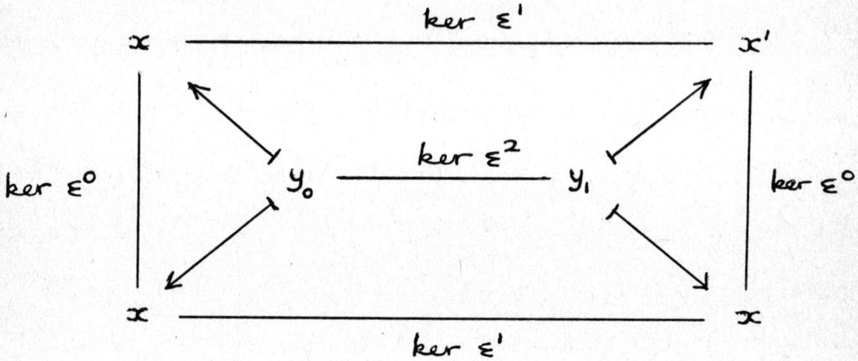

with $-y_0\rho = z' - z - y_1\rho$. Since $x\varepsilon^0 = x'\varepsilon^0$, $x'\varepsilon^1 = x\varepsilon^1 = x\varepsilon^1\delta\varepsilon^1$, and $x\varepsilon^1\delta\varepsilon^0 = x\varepsilon^1$, $(x, x', x\varepsilon^1\delta) \in \ker((\varepsilon^0, \varepsilon^1) : RG^2 \Longrightarrow RG)$, so $\exists\, y \in RG^3$. $y\varepsilon^0 = x$, $y\varepsilon^1 = x'$, $y\varepsilon^2 = x\varepsilon^1\delta$. Thus $(x,z)\ U\ (x', z+y\rho)$. But

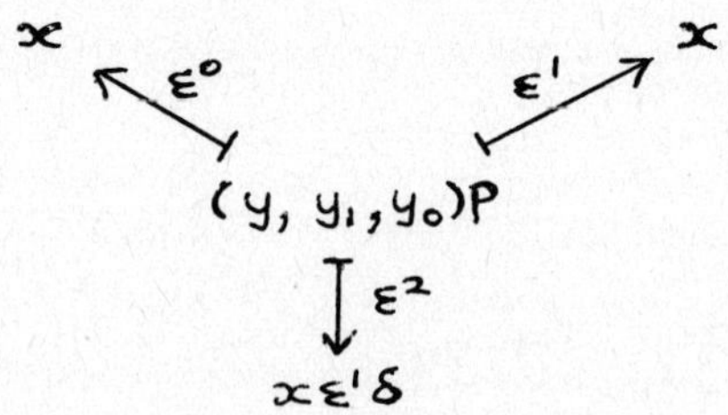

,

as is routinely checked, so $(x,z)\ U\ (x,z+y\rho-y_1\rho+y_0\rho) = (x,z+z-z'+y\rho)$. By 642, $z+z-z'+y\rho = z$, so $(x,z)\ U\ (x',z+y\rho) = (x',z')$. Thus (C4) for $(E^1|E^0)$ holds. Corollary 224 then shows that E^1 centralises E^0 by means of the centreing congruence $(E^1|E^0)$. By the symmetry of centralising, E^0 centralises E^1 by means of some $(E^0|E^1)$.

Suppose a congruence strictly containing E^0 centralises E^1. Define a relation φ on RG by $y\ \varphi\ y' \Leftrightarrow \exists\ (x,z)\ ,\ (x',z') \in \underline{A}\ .\ x\varepsilon^0 = y\ ,\ x'\varepsilon^0 = y'$, and $(x,z)^U\ F\ (x',z')^U$. If $y \in RG$, $(y\delta,0)^U\ F\ (y\delta,0)^U$, and $y\delta\varepsilon^0 = y$, so $y\ \varphi\ y$. Clearly $\varphi \lesssim {}^2(RG)$, and thus φ is a congruence on RG. Since $F > E^0$, $\varphi > \widehat{RG}$. Define a relation $(\varphi|\ker\varepsilon)$ on $\ker\varepsilon$ by (y_{00},y_{01}) $(\varphi|\ker\varepsilon)\ (y_{10},y_{11})$ iff $\exists\ (x_{ij},z_{ij}) \in \underline{A}\ .\ x_{ij}\varepsilon^0 = y_{ij}$ for $i, j \in \{0,1\}$ and $((x_{00},z_{00})^U,(x_{01},z_{01})^U)\ (F|E^1)\ ((x_{10},z_{10})^U,(x_{11},z_{11})^U)$. Note that $(x_{i0},z_{i0})^U\ E^1\ (x_{i1},z_{i1})^U$ implies that $y_{i0}\varepsilon = x_{i0}\varepsilon^0\varepsilon = x_{i0}\varepsilon^1\varepsilon = x_{i1}\varepsilon^1\varepsilon = x_{i1}\varepsilon^0\varepsilon = y_{i1}\varepsilon$ for $i = 0,1$. Clearly $(\varphi|\ker\varepsilon) \lesssim {}^2\ker\varepsilon$. Let $(y_0,y_1) \in \ker\varepsilon$. By Proposition 612 $\exists\ (x_0,x_1) \in \ker\varepsilon^1$. $y_i = x_i\varepsilon^0$, $i = 0,1$. Then $((x_0,0)^U,(x_1,0)^U)\ (F|E^1)$ $((x_0,0)^U,(x_1,0)^U)$, so $(y_0,y_1)\ (\varphi|\ker\varepsilon)\ (y_0,y_1)$. Thus $(\varphi|\ker\varepsilon)$ is a congruence on $\ker\varepsilon$.

Suppose $(y,y)\ (\varphi|\ker\varepsilon)\ (y_0,y_1)$, say $((x,z)^U,(x',z')^U)\ (F|E^1)\ ((x_0,z_0)^U,(x_1,z_1)^U)$ with $x\varepsilon^0 = y = x'\varepsilon^0$,

$x_i\varepsilon^0 = y_i$, $i = 0,1$. Now $(x,x') \in \ker\varepsilon^0 \cap \ker\varepsilon^1$ implies, by 642, that $\exists\, z''$. $(x',z')\ U\ (x,z'')$. Then $((x_0,z_0)^U,(x_1,z_1)^U)\ (F|E^1)$ $((x\ ,z)^U,(x\ ,z'')^U)\ (F|E^1)\ ((x_0,z_0)^U,(x_0,(z'',z,z_0)P)^U)$. $\Rightarrow (x_1,z_1)\ U$ $(x\ ,(z'',z,z_0)P)$. $\Rightarrow y_1 = x_1\varepsilon^0 = x_0\varepsilon^0 = y_0$. (C4) for $(\varphi|\ker\varepsilon)$ is verified. Thus by Corollary 224 φ , a congruence strictly containing $\widehat{RG}$, centralises $\ker\varepsilon$, contrary to the hypothesis on R in $\underline{\underline{T}}$. The conclusion is that $E^0 = \eta(E^1)$ on A .

Let $B^2 = E^1\text{nat}(E^0|E^1)$, $B^1 = RG$, $B^0 = R$. Define $e^i : B^2 \to B^1$; $((x_0,z_0)^U,(x_1,z_1)^U)\text{nat}(E^1|E^0) \mapsto x_i\varepsilon^0$ for $i = 0,1$; these are clearly well-defined. Let $e = \varepsilon : B^1 = RG \to B^0 = R$. Let $B^3 = \ker((e^0,e^1) : B^2 \Rrightarrow B^1)$, and let B^* be the simplicial object $B^3 \Rrightarrow B^2 \rightrightarrows B^1 \to B^0$. Since $E^0 = \eta(E^1)$ on A , it follows from Proposition 229 that $\ker(e^0 : B^2 \to B^1) = \eta(\ker(e^1 : B^2 \to B^1))$. Clearly $e : B^1 \to B^0$ surjects. Surjacency of B^* on B^1 follows from Proposition 612. Thus B^* is a seeded simplicial object.

Let $p^2 : RG^2 \to B^2$; $x \mapsto ((x,0)^U,(x\varepsilon^1\delta,0)^U)^{(E^0|E^1)}$. Let $p^1 = 1_{RG} : RG \to B^1$, $p^0 = 1_R : R \to B^0$. For $x \in RG$, $xp^1e = x\varepsilon = x\varepsilon p^0$. For $x \in RG^2$, $xp^2e^0 = x\varepsilon^0 = x\varepsilon^0p^1$ and $xp^2e^1 = x\varepsilon^1\delta\varepsilon^0 = x\varepsilon^1 = x\varepsilon^1p^1$. Define $p^3 : RG^3 \to B^3$; $x \mapsto (x\varepsilon^0p^2,x\varepsilon^1p^2,x\varepsilon^2p^2)$. Clearly for $x \in RG^3$, $xp^3e^i = x\varepsilon^ip^2$ for $i = 0,1,2$. Thus $p^* = (p^0,p^1,p^2,p^3,)$ is a simplicial map from RG^* to B^* . The module M grown by B* becomes an R-module. With the usual notation, $C =$ $\{ ((x,z)^U,(x',z')^U)\mathrm{nat}(E^0|E^1) \mid (x,x') \in \ker\varepsilon^0 \cap \ker\varepsilon^1 \} =$ $\{ ((x,z)^U,(x,z'')^U)\mathrm{nat}(E^0|E^1) \}$ by 642 . Then there is an isomorphism of R-modules $\theta : M = C^V \to Z \ ; \ (((x,z)^U,(x,z'')^U)\mathrm{nat}(E^0|E^1))^V \mapsto z''-z$.

Let x be in RG . Then

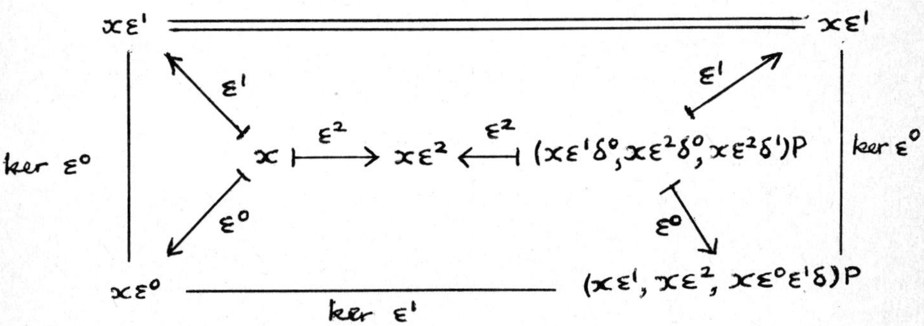

and so since $\rho \in \mathrm{Der}_0(RG^3,Z)$, $((x\varepsilon^0,x\rho)^U,(x\varepsilon^1,0)^U)\ (E^1|E^0)$ $(((x\varepsilon^1,x\varepsilon^2,x\varepsilon^0\varepsilon^1\delta)P,0)^U,(x\varepsilon^1,0)^U)$. Thus $(x\varepsilon^0,x\rho)\ U$ $((x\varepsilon^1,x\varepsilon^2,x\varepsilon^0\varepsilon^1\delta)P,0)$, whence $((x\varepsilon^0,x\rho)^U,(x\varepsilon^0\varepsilon^1\delta,0)^U)\ (E^0|E^1)$ $((x\varepsilon^1,0)^U,(x\varepsilon^2,0)^U)\ (E^0|E^1)\ ((x\varepsilon^0,0)^U,((x\varepsilon^0,x\varepsilon^1,x\varepsilon^2)P,0)^U)$. This implies that $(((x\varepsilon^0,x\varepsilon^1,x\varepsilon^2)P,0)^U,(x\varepsilon^0\varepsilon^1\delta,0)^U)\ (E^1|E^0)$ $((x\varepsilon^0,0)^U,(x\varepsilon^0,x\rho)^U)$. But both sides of this relation are in $E^0 \cap E^1$, and so it can be written as $(((x\varepsilon^0,x\varepsilon^1,x\varepsilon^2)P,0)^U,(x\varepsilon^0\varepsilon^1\delta,0)^U)$ $(E^0 \circ E^1|E^0 \cap E^1)\ ((x\varepsilon^0,0)^U,(x\varepsilon^0,x\rho)^U)$. Thus $x\rho^3P =$ $((x\varepsilon^0,0)^U,(x\varepsilon^0,x\rho)^U)^{(E^0|E^1)}$, and $x\rho^3P^V$ corresponds to $x\rho$ in Z under the isomorphism $\theta : C^V \to Z$ of R-modules, as required.]

NOTES

The extension theory given in this chapter generalises that given by Gruenberg for groups [Gr, Chapter 5], Barr for commutative algebras [Ec, pp. 357 - 375] and Orzech for "categories of interest" [Or]. It is modelled quite closely on these latter two. Seeded simplicial objects correspond to Gruenberg's (G:A)-cores [Gr, p. 73] and Barr's class $\mathbb{E}$ [Ec, p. 359]. The simplicial object RG* may well be too large for practical purposes, although it can easily be replaced by something smaller. Its big advantage in the current setting is its cleanliness. Duskin [Du] has given interpretations of all the groups $H^n(R,M)$ for an R-module M in great generality, corresponding for $n = 1$ and 2 to Theorems 632 and 641.

The approach in section 6.1 was to develop the minimum of necessary machinery as directly as possible. Fancier theoretical explanations may be found in [Du], [Ec], and their references.

R E F E R E N C E S

[AC] A. Almeida Costa, "Cours d'Algèbre Générale", Fundação Calouste Gulbenkian, Lisboa, 1969.

[Bi] G. Birkhoff, "Lattice Theory" 3rd. ed., American Mathematical Society, Providence R. I., 1967.

[Ca] B. R. Caine, A characterisation of some equationally complete varieties of quasigroups, preprint.

[Co] P. M. Cohn, "Universal Algebra", Harper and Row, New York, 1965.

[Du] J. W. Duskin, Simplicial methods and the interpretation of "triple" cohomology, Mem. Am. Math. Soc. No. 163, 1975.

[Ec] B. Eckmann (ed.), "Seminar on Triples and Categorical Homology Theory", Springer Lecture Notes in Math. (No. 80), Springer-Verlag, Berlin-Heidelberg-New York, 1969.

[FP] A. L. Foster and A. F. Pixley, Semi-categorical algebras II, Math. Z. 85 (1964), 169 - 184.

[Fr] P. Freyd, Algebra valued functors in general and tensor products in particular, Colloq. Math. 14 (1966), 89 - 106.

[Go] A. W. Goldie, On direct decompositions I, II, Proc. Cambridge Philos. Soc. 48 (1952), 1 - 34.

[Gr] K. W. Gruenberg, "Cohomological Topics in Group Theory", Springer Lecture Notes in Math. (No. 143), Springer-Verlag,

Berlin-Heidelberg-New York, 1970.

[Hu] B. Huppert, "Endliche Gruppen I", Springer-Verlag, Berlin-Heidelberg-New York, 1967.

[Ja] N. Jacobson, "Structure of Rings", A.M.S. Colloq. Publ. Vol. XXXVII, Providence R.I., 1956.

[JT] B. Jónsson and A. Tarski, "Direct Decompositions of Finite Algebraic Systems", Notre Dame Mathematical Lectures, Notre Dame, Indiana, 1947.

[Ma] A. I. Mal'cev, K obščeĭ teorii algebraičeskih sistem, Mat. Sb. N. S. 35 (77) (1954), 3 - 20.

[MH] Yu. I. Manin (trans. M. Hazewinkel), "Cubic Forms", North-Holland, Amsterdam, 1974.

[Mi] Ju. I. Manin, "Kubičeskie Formy", Nauka, Moskva, 1972.

[ML] S. MacLane, "Categories for the Working Mathematician", Springer-Verlag, New York-Heidelberg-Berlin, 1971.

[Mu] D. C. Murdoch, Structure of abelian quasi-groups, Trans. Amer. Math. Soc. 49 (1941), 392 - 409.

[Or] G. G. Orzech, "Obstruction Theory in Algebraic Categories", Ph. D. Thesis, Urbana, Illinois, 1970.

[PA] H. (Popova) Alderson, The structure of the logarithmetics of finite plain quasigroups, J. Alg. 31 (1974), 1 - 9.

[Wr] G. C. Wraith, "Algebraic Theories", Aarhus Lecture Notes Series (No. 22), Aarhus, 1970.

A L P H A B E T I C A L   I N D E X

INDEX OF (NON-ALPHABETICAL) NOTATION

Vol. 457: Fractional Calculus and Its Applications. Proceedings [illegible] 1974. Edited by B. Ross. VI, 381 pages. 1975.

Vol. 458: P. Walters, Ergodic Theory – Introductory Lectures. VI, 198 pages. 1975.

Vol. 459: Fourier Integral Operators and Partial Differential Equations. Proceedings 1974. Edited by J. Chazarain. VI, 372 pages. 1975.

Vol. 460: O. Loos, Jordan Pairs. XVI, 218 pages. 1975.

Vol. 461: Computational Mechanics. Proceedings 1974. Edited by J. T. Oden. VII, 328 pages. 1975.

Vol. 462: P. Gérardin, Construction de Séries Discrètes p-adiques. »Sur les séries discrètes non ramifiées des groupes réductifs déployés p-adiques«. III, 180 pages. 1975.

Vol. 463: H.-H. Kuo, Gaussian Measures in Banach Spaces. VI, 224 pages. 1975.

Vol. 464: C. Rockland, Hypoellipticity and Eigenvalue Asymptotics. III, 171 pages. 1975.

Vol. 465: Séminaire de Probabilités IX. Proceedings 1973/74. Edité par P. A. Meyer. IV, 589 pages. 1975.

Vol. 466: Non-Commutative Harmonic Analysis. Proceedings 1974. Edited by J. Carmona, J. Dixmier and M. Vergne. VI, 231 pages. 1975.

Vol. 467: M. R. Essén, The Cos $\pi\lambda$ Theorem. With a paper by Christer Borell. VII, 112 pages. 1975.

Vol. 468: Dynamical Systems – Warwick 1974. Proceedings 1973/74. Edited by A. Manning. X, 405 pages. 1975.

Vol. 469: E. Binz, Continuous Convergence on C(X). IX, 140 pages. 1975.

Vol. 470: R. Bowen, Equilibrium States and the Ergodic Theory of Anosov Diffeomorphisms. III, 108 pages. 1975.

Vol. 471: R. S. Hamilton, Harmonic Maps of Manifolds with Boundary. III, 168 pages. 1975.

Vol. 472: Probability-Winter School. Proceedings 1975. Edited by Z. Ciesielski, K. Urbanik, and W. A. Woyczyński. VI, 283 pages. 1975.

Vol. 473: D. Burghelea, R. Lashof, and M. Rothenberg, Groups of Automorphisms of Manifolds. (with an appendix by E. Pedersen) VII, 156 pages. 1975.

Vol. 474: Séminaire Pierre Lelong (Analyse) Année 1973/74. Edité par P. Lelong. VI, 182 pages. 1975.

Vol. 475: Répartition Modulo 1. Actes du Colloque de Marseille-Luminy, 4 au 7 Juin 1974. Edité par G. Rauzy. V, 258 pages. 1975. 1975.

Vol. 476: Modular Functions of One Variable IV. Proceedings 1972. Edited by B. J. Birch and W. Kuyk. V, 151 pages. 1975.

Vol. 477: Optimization and Optimal Control. Proceedings 1974. Edited by R. Bulirsch, W. Oettli, and J. Stoer. VII, 294 pages. 1975.

Vol. 478: G. Schober, Univalent Functions – Selected Topics. V, 200 pages. 1975.

Vol. 479: S. D. Fisher and J. W. Jerome, Minimum Norm Extremals in Function Spaces. With Applications to Classical and Modern Analysis. VIII, 209 pages. 1975.

Vol. 480: X. M. Fernique, J. P. Conze et J. Gani, Ecole d'Eté de Probabilités de Saint-Flour IV-1974. Edité par P.-L. Hennequin. XI, 293 pages. 1975.

Vol. 481: M. de Guzmán, Differentiation of Integrals in R^n. XII, 226 pages. 1975.

Vol. 482: Fonctions de Plusieurs Variables Complexes II. Séminaire François Norguet 1974–1975. IX, 367 pages. 1975.

Vol. 483: R. D. M. Accola, Riemann Surfaces, Theta Functions, and Abelian Automorphisms Groups. III, 105 pages. 1975.

Vol. 484: Differential Topology and Geometry. Proceedings 1974. Edited by G. P. Joubert, R. P. Moussu, and R. H. Roussarie. IX, 287 pages. 1975.

[illegible] Geometry of Banach Spaces – Selected Topics. XI, 282 pages. 1975.

Vol. 486: S. Stratila and D. Voiculescu, Representations of AF-Algebras and of the Group U (∞). IX, 169 pages. 1975.

Vol. 487: H. M. Reimann und T. Rychener, Funktionen beschränkter mittlerer Oszillation. VI, 141 Seiten. 1975.

Vol. 488: Representations of Algebras, Ottawa 1974. Proceedings 1974. Edited by V. Dlab and P. Gabriel. XII, 378 pages. 1975.

Vol. 489: J. Bair and R. Fourneau, Etude Géométrique des Espaces Vectoriels. Une Introduction. VII, 185 pages. 1975.

Vol. 490: The Geometry of Metric and Linear Spaces. Proceedings 1974. Edited by L. M. Kelly. X, 244 pages. 1975.

Vol. 491: K. A. Broughan, Invariants for Real-Generated Uniform Topological and Algebraic Categories. X, 197 pages. 1975.

Vol. 492: [illegible] Carol Karp. Edited by D. W. Kueker. VI, 206 pages. 1975.

Vol. 493: F. W. Kamber and P. Tondeur, Foliated Bundles and Characteristic Classes. XIII, 208 pages. 1975.

Vol. 494: [illegible] Potential Resolvent [illegible]

Vol. 495: [illegible] Groups II. V, [illegible] pages. 1975.

Vol. 496: L. H. Hodgkin and V. P. Snaith, Topics in K-Theory. Two Independent Contributions. III, 294 pages. 1975.

Vol. 497: Analyse Harmonique sur les Groupes de Lie. Proceedings 1973–75. Edité par P. Eymard et al. VI, 710 pages. 1975.

Vol. 498: Model Theory and Algebra. A Memorial Tribute to Abraham Robinson. Edited by D. H. Saracino and V. B. Weispfenning. X, 463 pages. 1975.

Vol. 499: Logic Conference, Kiel 1974. Proceedings. Edited by G. H. Müller, A. Oberschelp, and K. Potthoff. V, 651 pages 1975.

Vol. 500: Proof Theory Symposion, Kiel 1974. Proceedings. Edited by J. Diller and G. H. Müller. VIII, 383 pages. 1975.

Vol. 501: Spline Functions, Karlsruhe 1975. Proceedings. Edited by K. Böhmer, G. Meinardus, and W. Schempp. VI, 421 pages. 1976.

Vol. 502: János Galambos, Representations of Real Numbers by Infinite Series. VI, 146 pages. 1976.

Vol. 503: Applications of Methods of Functional Analy[illegible] Problems in Mechanics. Proceedings 1975. Edited by P. G[illegible]ain and B. Nayroles. XIX, 531 pages. 1976.

Vol. 504: S. Lang and H. F. Trotter, Frobenius [illegible]butions in GL_2-Extensions. III, 274 pages. 1976.

Vol. 505: Advances in Complex Function Theory. Proceedings 1973/74. Edited by W. E. Kirwan and L. Zalcman. VIII, 203 pages. 1976.

Vol. 506: Numerical Analysis, Dundee 1975. Proceedings. Edited by G. A. Watson. X, 201 pages. 1976.

Vol. 507: M. C. Reed, Abstract Non-Linear Wave Equations. VI, 128 pages. 1976.

Vol. 508: E. Seneta, Regularly Varying Functions. [illegible] pages. 1976.

Vol. 509: D. E. Blair, Contact Manifolds in Rie[illegible] Geometry. VI, 146 pages. 1976.

Vol. 510: V. Poènaru, Singularités C^∞ en Pré[illegible] Symétrie. V, 174 pages. 1976.

Vol. 511: Séminaire de Probabilités X. Proceedi[illegible]/75. Edité par P. A. Meyer. VI, 593 pages. 1976.

Vol. 512: Spaces of Analytic Functions, Kristians[illegible]way 1975. Proceedings. Edited by O. B. Bekken, B. K. Økse[illegible] A. Stray. VIII, 204 pages. 1976.

Vol. 513: R. B. Warfield, Jr. Nilpotent Groups. VIII, 1[illegible]ges. 1976.

Vol. 514: Séminaire Bourbaki vol. 1974/75. Exposé[illegible] – 470. IV, 276 pages. 1976.

Vol. 515: Bäcklund Transformations. Nashville, Ten[illegible]see 1974. Proceedings. Edited by R. M. Miura. VIII, 295 pages. 1976.